Current Topics in Microbiology and Immunology

131

Editors

The Molecular Biology of Baculoviruses

Edited by W. Doerfler and P. Böhm

With 44 Figures

Springer-Verlag
Berlin Heidelberg New York
London Paris Tokyo

Professor Dr. WALTER DOERFLER
Institut für Genetik
der Universität zu Köln
Weyertal 121
D-5000 Köln 41

PETRA BÖHM
Institut für Genetik
der Universität zu Köln
Weyertal 121
D-5000 Köln 41

Softcover reprint of the hardcover 1st edition 1986

Library of Congress Catalog Card Number 15-12910

ISBN-13:978-3-642-71591-4 eISBN-13:978-3-642-71589-1
DOI:10.1007/978-3-642-71589-1

Preface

The pathology caused by baculoviruses in insect populations was described centuries ago, notably in the larvae of insects such as the silkworm (*Bombyx mori*) which has been appreciated for the quality and beauty of its products. In the 1940s baculoviruses and their structure and physiology were intensively investigated, particularly by Bergold's group in Tübingen. The following decades saw excellent progress, laying a solid virological base for later investigations on the system. Further studies mushroomed in the 1970s with the advent of tissue culture systems for insect cells which eventually facilitated the molecular biological approach that came to the fore in the 1980s.

One of the reasons for pursuing research on the baculovirus system was the prospect of eventually using these viruses as insect pest control agents. While this practical aspect may appeal to many, molecular biologists had additional reasons to be interested in baculoviruses. Here was a large DNA viral genome, probably fraught with problems of replication and regulation that hopefully would open inroads into the molecular biology of interesting insect cell systems. In the days when genetechnology promises laurels, and after several virus systems had been skilfully exploited as highly efficient eukaryotic expression vectors, it came as no surprise that baculoviruses were also investigated in that respect. Indeed, the *Autographa californica* nuclear polyhedrosis virus became a good vector. Insect cells also seem to collaborate in modifying and processing the genetechnologically synthesized polypeptides. History went a full cycle, as it were, when it was demonstrated how man's faithful working horse, the silkworm, could now be exploited to reap the fruits of baculovirus genetechnology.

The objective of this volume is not to present an encyclopedia of baculoviruses, but rather to offer a compendium for the interested student of baculovirology with current contributions from laboratories which are actively involved in work on the molecular biology of these viruses. The following topics will be considered:

Structure and replication of baculoviruses, the structure of the viral genome and its expression as well as the regulation of baculovirus gene expression are the topics of several chapters. One of these chapters also summarizes the current information on foreign genes that have been expressed by the baculovirus vector system. Another chapter is devoted to the molecular biology of granulosis viruses. It has been recognized for some time that insect cells can be persistently infected with baculoviruses, and the present state of investigations on this topic is the subject of another chapter. Detailed analyses on virus-specific proteins, like the 64-kilodalton envelope protein and the inclusion body protein, the polyhedrin, will also be presented. Biochemical analyses on the structure and function of virus-encoded proteins will certainly constitute a very important area for further research.

It is hoped that the papers collected in this volume will be useful for newcomers to the field seeking to familiarize themselves with current research work, as well as for specialists seeking detailed information. These reviews do not intend to treat baculovirology comprehensively. There will be other sources of information to fulfill that request. I am aware of the shortcoming that certain important areas of molecular biology, e.g., the problem of baculovirus DNA replication, have hardly been touched upon. Frankly, there is not enough definitive information from recent work to warrant a full chapter. Future books on the molecular biology of baculoviruses will hopefully help to alleviate some of these unavoidable omissions.

We would like to thank the authors who have contributed to this volume and transgressed the deadline only minimally or not at all. It is a pleasure to thank Springer-Verlag, in particular Marga Botsch, for their help and excellent collaboration.

Cologne, Summer 1986

WALTER DOERFLER
PETRA BÖHM

Table of Contents

Indexed in Current Contents

List of Contributors

You will find the addresses at the beginning of the respective contribution

An Overview of the Structure and Replication of Baculoviruses

P. FAULKNER and E.B. CARSTENS

1 Introduction

Diseases of insects, in particular those of the honeybee and the silkworm, have been recognized since ancient times and were the subject of speculation in historical writings. According to BERGOLD (1953), MAESTRI and CORNELIA in 1856 established a correlation between the presence of crystalline occlusion bodies within tissues of insects and the symptoms of disease in silkworms. A similar pattern of disease was seen in forest insects in the early part of this century, and the infections became known as nuclear polyhedrosis diseases. General acceptance that the agent was viral in nature followed demonstrations that extracts of silkworm tissue from jaundiced larvae retained infectivity after passing through an ultrafilter (VON PROWASEK 1912; ACQUA 1918). In 1947 BERGOLD demonstrated that rod-shaped virions could be released from occulusion bodies in the presence of dilute alkali (BERGOLD 1947) and that they were enveloped and contained DNA. Since that time there has been a continuing scientific interest in furthering knowledge of the part these viruses play in natural control of insect populations. Detailed biochemical studies on the nature of the virus

Department of Microbiology and Immunology, Queen's University, Kingston, Ontario, K7L 3N6, Canada

Current Topics in Microbiology and Immunology, Vol. 131

and its molecular biology have accumulated in recent years. Much of the current thrust of research is directed to assessing the role of the viruses in integrated insect pest management programs and, at the molecular level, in attempts to engineer viruses of altered insect host range and virulence. Other research is directed to utilizing baculoviruses as expression vectors for production of non-baculovirus proteins (MILLER et al. 1982; 1983a).

The family Baculoviridae contains a single genus, *Baculovirus*, which is divided into three subgroups (MATHEWS 1982). (1) Subgroup A contains the nuclear polyhedrosis viruses (NPV). Nuclei of infected cells become filled with crystalline polyhedron-shaped occlusion bodies (OB) which are up to 15 μm in size and are easily seen in the light microscope. About 95% of the mass of the OB is made up of a crystalline protein lattice in which are embedded enveloped virus nucleocapsids. The viruses are dispersed throughout the OB either as separate enveloped nucleocapsids (SNPV) or as bundles of one to seven nucleocapsids within a common envelope (MNPV). (2) Subgroup B consists of the granulosis viruses (GV); here the OB are much smaller (about 0.5 μm) and are ellipsoidal in shape. They occlude a single enveloped virus particle. (3) The type virus in subgroup C is *Oryctes* baculovirus which infects the rhinoceros beetle. The virion is a rod-shaped, enveloped nucleocapsid which does not become occluded and is transmitted as the free particle.

Most investigations on gene expression of baculoviruses have been done using *Autographa californica* nuclear polyhedrosis virus (AcMNPV). Studies have also been carried out on other baculoviruses, particularly those which infect pest insects and result in agricultural crop or forest damage. Recent reviews on baculovirus structure and genome organization and expression are by HARRAP and PAYNE (1979), MILLER (1984), KELLY (1985), and COCHRAN et al. (1986).

2 The Infectious Cycle

Permissive cell lines (reviewed in KNUDSON and BUCKLEY 1977 and PRISTON 1985) have been used for over a decade to study the morphogenesis and biochemistry of replication of nuclear polyhedrosis viruses. Recently cell cultures have been developed which support a codling moth granulosis virus (NASER et al. 1984; MILTENBERGER et al. 1984) and two subgroup C nonoccluded baculoviruses: from the rhinoceros beetle (CRAWFORD and SHEEHAN 1985) and from *Heliothis zea* (RALSTON et al. 1981).

The replication of NPV in cell culture is biphasic and two phenotypes of the virion are generated (VOLKMAN et al. 1976) in the infectious cycle (reviewed in: GRANADOS 1980; FAULKNER 1981; VAUGHN and DOUGHERTY 1985). AcMNPV infection of cell cultures is initiated by the nonoccluded form of the virus (NOV). The virion, which is an enveloped nucleocapsid, becomes internalized by the process of adsorptive endocytosis (VOLKMAN and GOLDSMITH 1985). Uncoated nucleocapsids traverse the cytoplasm and either enter the nucleus or discharge the genome at nuclear pores. Virus entry appears to be an efficient process since nucleocapsids have been seen in the nucleus within 1 h of infection (BASSEMIR et al. 1983). Transcription and DNA replication occur within the nucleus and unenveloped nucleocapsids are observed within the first 12 h postinfection. Morphogenesis of the nucleocapsid apparently first involves assembly of capsids followed by addition of nucleoprotein to the capsid shells (BASSEMIR et al. 1983). In virus morphogenesis the nucleocapsids are first observed lying within patches of DNA-rich virogenic stroma which form in infected nuclei. Early in infection, newly formed unenveloped nucleocapsids traverse the cytoplasm and progeny NOV are released by budding at the plasma membrane (GRANADOS 1980; FAULKNER 1981). In the second phase of the infectious cycle budding of NOV appears to be reduced (VOLKMAN et al. 1976) and *de novo* membrane proliferation may occur within the nucleus (STOLTZ et al. 1973). Bundles of enveloped nucleocapsids accumulate in the nucleus and then become encapsulated within the matrix of the occlusion bodies. Occlusion of the newly assembled occluded virus (OV) is by crystallization of polyhedrin around the bundles (CHUNG et al. 1980).

In nature the two baculovirus phenotypes have different functions in the replication and transmission of the virus. Larvae of Lepidopteran insects become infected when baculovirus occlusion bodies are ingested along with the food. They dissolve at about pH 10.5 in gut juice and bundles of OV are released. The OV infect gut epithelial cells and virus replication takes place. Enveloped nucleocapsids of NOV phenotype are budded from the basal surface of gut cells into the hemocoel (ADAMS et al. 1977). A generalized dispersion of the virus follows, with many tissues becoming infected. Secondary rounds of replication occur and ultimately occlusion bodies form within nuclei in most tissues of the larvae (FAULKNER 1981).

3 Properties of the NOV and OV Phenotypes

Envelopes and capsids of NOV and OV differ in polypeptide (VOLKMAN 1983) and glycoprotein (STILES and WOOD 1983) composition but the genomes from both phenotypes have a similar DNA sequence (COCHRAN et al. 1982). A recent study of surface envelope proteins by lactoperoxidase iodination identified at least 12 polypeptides in AcMNPV NOV and 13 in the OV form. Eight of the NOV but none of the OV polypeptides were glycosylated (STILES and WOOD 1983). The phenotypes have differing infectivities according to the route of administration. Under experimental conditions OV have high infectivity *per os* but are of low infectivity if injected into the hemocoele. In contrast NOV are of low infectivity per os but highly infectious when injected (DOUGHERTY et al. 1975; KEDDIE and VOLKMAN 1985).

Envelopes of both phenotypes of baculoviruses can be removed by detergent treatment. The nucleocapsids are rod shaped and about 40 nm in diameter and 300 nm long. Further treatment of the nucleocapsids with dilute alkali or with detergent and high salt (WILSON and CONSIGLI 1985b) separates the capsids and virus cores. Capsids retain the cylindrical form of the nucleocapsid, and a lattice pattern of capsomers is seen in the EM (HARRAP 1972; BEATON and FILSHIE 1976). The predominant capsid polypeptide is 44K in size (SUMMERS and SMITH 1978), but up to eight minor polypeptides may also be present (KELLY 1985). Purified capsids of a GV virus have protein kinase activity (WILSON and CONSIGLI 1985b). The enzyme requires Mn^{2+} for activation and can phosphorylate the 13K virus core DNA-binding protein. Similar basic proteins have been isolated from NPV and other GV baculoviruses (TWEETEN et al. 1981; KELLY et al. 1982). It has been postulated that during cellular infection the 13K protein is phosphorylated by a capsid protein kinase and that virus DNA is released from the core complex. WILSON and CONSIGLI (1985) have proposed that in GV infections nucleocapsids become aligned with nuclear pores and protein kinase is activated. The action of the enzyme would then be to allow for release of genomic DNA into the nucleus to initiate infection.

4 The Baculovirus Genome

The genome of the baculoviruses is a closed circular superhelical molecule of double-stranded DNA. Genomes ranging in size from 60 to 250 kbp have been reported (MILLER 1984), but the median is about 100 kbp. The general conformation was established based on biophysical properties, e.g., behavior of virion DNA in density gradients in the presence of intercalating agents and by direct observation in the electron microscope of DNA released from virus particles. (Some recent reviews are: HARRAP and PAYNE 1979; FAULKNER 1981; MILLER 1984; KELLY 1985.) The degree of supercoiling has not been accurately determined because most DNA extracted from virions is in the form of nicked (relaxed) circles. The virion vDNA is believed to be in tight association with a strongly basic polypeptide of approximately 13K molecular weight which stabilizes the DNA within the nucleocapsid (TWEETEN et al. 1981b; KELLY et al. 1982).

4.1 The Physical Map

Development of a physical map provided an important step in elucidating the gene order in baculoviruses. In a physical map, DNA fragments generated with class II restriction endonucleases (REN) are ordered on a linear or circular scale. By convention the scale is usually divided into 100 units and the actual location of REN fragments or specific sequences are reported as a range within the scale of 0–100 map units. Strategies used in developing physical maps of baculoviruses have been reviewed by MILLER (1984) and by COCHRAN et al. (1986).

The first physical maps of the baculovirus genome were with variants of AcMNPV (E-2, SMITH and SUMMERS 1979, VLAK 1980; L-1, MILLER and DAWES 1979; HR-3, COCHRAN et al. 1982; and D, LUBBERT et al. 1981). The circular nature of the genome was confirmed in these studies and estimates of the molecular weight of the genome were obtained based on summation of the sizes of the REN fragments. The physical map was oriented and the 0/100 locus defined in keeping with a consensus proposal among baculovirologists (VLAK and SMITH 1982). The map originated and ended at the *Eco*R1 B/I junction of the E-2, L-1, and HR-3 variants and the B/H junction of the D variant. The rationale for using the *Eco*R1 B/I junction for orientation was based on studies which located the polyhedrin gene within the *Eco*R1-I fragment of the E-2 variant (SUMMERS et al. 1980). The AcMNPV map has been utilized in studies on gene composition and expression. Portions of it have been finemapped and sequenced (HOOFT VAN IDDEKINGE et al. 1983; KUZIO et al. 1984).

In the past few years physical maps have been published for other baculoviruses and their strain variants. The list includes the AcMNPV variants: *Spodoptera exempta* MNPV (BROWN et al. 1985), *Galleria melonella* MNPV, and *Trichoplusia ni* MNPV (SMITH and SUMMERS 1979). In addition physical maps have been published on MNPV isolated from *Choristoneura fumiferana*, CfMNPV (spruce budworm) (ARIF et al. 1984), *Orgyia pseudotsugata*, OpMNPV (tussock moth) (LEISY et al. 1984), *Spodoptera frugiperda*, SfMNPV (MARUNIAK et al. 1984), and *Mamestra brassicae* (WIEGERS and VLAK 1984). Among SNPV, genomes of baculoviruses from *Heliothis zea* HzSNPV (KNELL and SUMMERS 1984), assorted Plusiine hosts (BILIMORIA 1983), and *Orgyia pseudotsugata* have been subjected to REN analysis. GVs from *Plodia interpunctella* and *Pieris rapae* (TWEETEN et al. 1980a) and the genome of the nonoccluded group C *Oryctes* baculovirus (CRAWFORD et al. 1985) have also been analyzed.

4.2 Variants and Strains Related to AcMNPV

As attention focussed on details of the physical map of AcMNPV it became apparent that cloned isolates of the virus often differed from the prototype strain in the fine structure of the genome. The differences in DNA sequence may be reduplication of portions of the genome, deletions of sequences, and base substitutions which may change the REN pattern of the isolate. Variations of genome structure have been observed in field isolates of AcMNPV (LEE and MILLER 1978; SMITH and SUMMERS 1979; TJIA et al. 1979; KNELL and SUMMERS 1981), *Cydia pomella* GV (HARVEY and VOLKMAN 1983), and other baculoviruses. They have also been induced in cell culture, especially following serial passage of AcMNPV (FRASER and HINK 1982; FRASER et al. 1985) or by subjecting the replicating virus to mutagens during replication (ERLANDSON et al. 1984). Instances of insertion of host cell DNA including the insertion of transposon-like elements have also been reported when the virus is grown in cell culture (MILLER and MILLER 1982; FRASER et al. 1983). A comprehensive physical map describing the variations found in numerous isolates of AcMNPV is given in the review by MILLER (1984).

Observations on variations in the physical map of AcMNPV have raised questions on the meaning of species, strain, and variant in the taxonomy of the baculoviruses. Until recently it has been the practice to name baculoviruses after the insect from which they were originally isolated (TINSLEY and HARRAP

1978); thus, NPVs derived from *Autographa californica*, *Galleria melonella*, *Rachoplusia ou*, and *Trichoplusia ni* were treated as separate species. It has been shown that these viruses differ only slightly in biological properties and their genomes have only minor variations in their physical maps (SMITH and SUMMERS 1979; SUMMERS et al. 1980). The use of Southern blotting and dot-blotting techniques has helped delineate the close similarity between many AcMNPV variants (SMITH and SUMMERS 1982). Investigations of such biological properties as host range and virulence have also often borne out these close relationships (HARVEY and VOLKMAN 1983). The taxonomic problem of distinguishing betwen baculovirus strains and variants remains. A useful approach is taken by COCHRAN et al. (1986), who point out that variants may be considered to be isolates derived from the same insect host, while strains are isolates derived from other insect species. Thus the four AcMNPV variants, E-2, HR-3, L-1, and D, were all isolated and cloned from early passage stocks of field collections of *Autographa californica*. However, there are only minor differences in the physical maps and REN profiles of AcMNPV, TnMNPV, and Se-25MNPV (BROWN et al. 1984), but these would be considered as strains since they were isolated from different insect species. In practice investigators need only be aware that divergences from a "prototype" sequence are often observed in a working strain. Nowadays most molecular genetic studies involve fine mapping of a locus within the genome and precise definitions of gene structure are given in each investigation.

4.3 Reiterated DNA Sequences Within Genomes

The genome of AcMNPV possesses five regions (HR 1–5) which show DNA homology when examined by Southern blotting under stringent hybridization conditions. The homologous regions are 350–500 bp in size and are rich in *Eco*R1 sites (COCHRAN and FAULKNER 1983). Altogether they represent about 3% of the total genome size. The homologous regions were detected by Southern blot and cross-blot hybridization and regions of homology were within the *Hind*III-A, -K, -L, -Q fragments. The *Hind*-L, -Q fragments were digested with *Eco*R1 and cloned into the plasmid pBR322. The cloned "minifragments" were shown to hybridize to all five homologous regions of the AcMNPV genome.

Regions of DNA homology have also been found in *Choristoneura fumiferana* NPV (ARIF and DOERFLER 1984) and possibly in *Orgyia pseudotsugata* NPV (LEISY et al. 1984). In addition, the subgroup-C *Oryctes* baculovirus contains four regions of sequence homology (CRAWFORD et al. 1985).

Intragenic reiterated sequences from a given baculovirus genome do not seem to be conserved in DNAs of other baculovirus species. Thus, although extensive homology has been noted between the genomes of AcMNPV and CfMNPV (ARIF and BROWN 1975) the homologous regions from the two viruses do not share the same sets of reiterated sequences (COCHRAN et al. 1986). The biological significance of the homologous repeat regions is largely unknown. The HR-5 region of AcMNPV is not essential for replication of the virus in the cell culture since a deletion mutant lacking it grew normally (J. KUZIO, personal communication). HR-5 is flanked by a transcriptionally active locus in the AcMNPV genome (LUBBERT and DOERFLER 1984b; FRIESEN and MILLER 1985; RANKIN et al. 1986) and may possibly have a role to play in regulating adjacent genes.

In addition to the homologous regions the genomes of baculoviruses may also generate insertions, reduplications of sequences, and transposon-like elements with flanking LTRs following continuous passage in cell culture (MILLER and MILLER 1982; FRASER and HINK 1982; FRASER et al. 1983). Should insertional mutagenesis of transposons occur, major effects on gene regulation within the virus could be anticipated. There are no reports of acquisition of host DNA sequences in wild-type isolates of baculoviruses.

4.4 Conservation of DNA Sequences Between Genomes

There have been several reports of intergenic homology among baculoviruses. Homologies were detected among Lepidopteran NPVs (JEWELL and MILLER 1980) and, using dot-blot hybridization at low and high stringenices, between a large group of MNPVs, SNPVs, GVs and a subgroup-C baculovirus (SMITH and SUMMERS 1982). The polyhedrin gene and a region encompassing the *p10* gene are highly conserved among NPVs infecting Lepidoptera.

Recent findings of homologies between large segments of the genomes of AcMNPV and OpMNPV (LEISY et al. 1984), AcMNPV and CfMNPV (ARIF et al. 1985), and AcMNPV and HzSNPV (KNELL and SUMMERS 1984) have indicated that the three viruses are colinear with respect to their gene distribution. The data point to strong evolutionary pressure that has preserved the structural organization of baculovirus genomes. Other evidence for conservation of DNA structure among the baculoviruses comes from studies made by Rohrmann and colleagues on the evolutionary history of the gene coding for the major occlusion body polyhedrin. Predictions of base sequence derived from N-terminal analyses of polyhedrins from baculoviruses derived from Lepidoptera, Hymenoptera, and Diptera, indicated major conservation of the gene through the evolutionary history of the viruses (ROHRMANN et al. 1981).

5 Replication of Baculovirus DNA

Baculoviruses have the largest circular virus genomes known. The genomes of several baculoviruses have been examined extensively and features which regulate transcription (ROHEL and FAULKNER 1984; ERLANDSON and CARSTENS 1983; ERLANDSON et al. 1985) and translation (ESCHE et al. 1982; VLAK et al. 1981) have been mapped in a few instances. However, little is known at present about the biochemical mechanisms by which the genome is replicated. Since purified viral DNA is infectious (CARSTENS et al. 1980; POTTER and MILLER 1980; BURAND et al. 1980), the capsid-free genome must be in a conformation ready for transcription as soon as it enters the cell nucleus. Whether this DNA associates with host cell proteins prior to transcription has yet to be determined but it is likely that nucleosomes assemble to protect the DNA from nucleases prior to the initiation of newly synthesized virus DNA (L.K. MILLER, personal communication). At least one report has suggested that the input DNA can be reincorporated in progeny virions (DOBOS and COCHRAN 1980). In cell cultures vDNA synthesis begins about 6 h postinfection and proceeds for at least another 12 h (TJIA et al. 1979; ERLANDSON and CARSTENS 1983). There is probably a decrease in the rate of vDNA synthesis after 20 h postinfection (TJIA et al. 1979) although the kinetics have not been well studied. A specific endonuclease activity has been detected in infected cells. The enzyme degrades cellular but not virus DNA (KELLY and WANG 1981; BROWN et al. 1979). Its presence could explain the shut-off of transcription and translation of cellular genes observed at late times after infection (CARSTENS et al. 1979; VLAK et al. 1981; ESCHE et al. 1982; ROHEL et al. 1983; LUBBERT and DOERFLER 1984a). The identification of a virus gene product responsible for this inhibition of cellular processes has great potential for developing a strain of virus with enhanced virulence by increasing either the level of gene expression or the length of time that it is synthesized.

Prior synthesis of virus polypeptides is necessary for the initiation of vDNA replication (KELLY 1982; GORDON and CARSTENS 1984; ERLANDSON et al. 1985) although their identification and the role of these proteins in virus infection has not yet been determined. A partial inhibition of vDNA synthesis was seen when infected cells were treated with drugs such as cycloheximide, actinomycin D, and cytosine arabinoside which inhibit protein and DNA synthesis (KELLY and LESCOTT 1976). However, some vDNA synthesis continued even in the presence of high drug concentrations. A similar effect on the level of protein synthesis has also been observed (DOBOS and COCHRAN 1980). More sensitive techniques of analysis such as Southern and dot-blot hybridizations need to be utilized to clarify relationships between protein synthesis and vDNA replication. Such studies could, for example, investigate whether virus transcripts are produced in the absence of protein synthesis and whether inhibition of protein synthesis completely blocks DNA replication.

Another approach to study baculovirus DNA replication has utilized temperature-sensitive mutants which depress the synthesis of vDNA at the nonpermissive temperature. Two such AcMNPV mutants, ts8 and tsB821, have been described (BROWN et al. 1979; GORDON and CARSTENS 1984; MILLER et al. 1983b). Both grow normally at the permissive temperature but fail to synthesize significant levels of vDNA at the restricted temperature. Five virus-induced proteins have been detected in mutant-infected cells at the nonpermissive temperature. These appear to be early proteins which are normally produced prior to initiation of DNA replication. Both mutants have been mapped by marker rescue to different regions of the genome. The ts8 mutation maps within 60.1 and 62 map units (GORDON and CARSTENS 1984) and the tsB821 lesion maps between 90.7 and 1.9 map units (MILLER et al. 1983b). The 60.1–62 map unit region codes for a 4.6-kb transcript in Wt-infected cells (ERLANDSON et al. 1985) and may encode the protein defective in ts8. No specific RNA has been associated yet with tsB821 and no polypeptide has been assigned to the altered sites in either mutant. Recently, an immediate early gene product required for the initiation of transcription from another other early promoter (for a p39 polypeptide) has been mapped between 95 and 97.5 map units (GUARINO and SUMMERS 1986). It remains to be determined whether this is the gene which is altered in tsB821.

Only one report has described a possible origin of replication on the genome of a baculovirus (BLINOV et al. 1984). Location of the tentative origin for GmMNPV (a variant of AcMNPV) was based on the ability of DNA from insect cells previously transfected with plasmids carrying vDNA fragments to transform *E. coli* competent cells. The authors found that the GmMNPV *Bam*HI fragment H carried an origin of replication. By comparing their fragment designations with published maps (SMITH and SUMMERS 1979), it is likely that the region of interest is homologous with the AcMNPV *Bam*HI G fragment. These results await confirmation. It would perhaps be coincidental that this region contains a legitimate baculovirus origin because other work suggests that virus-specific protein synthesis is required prior to the initiation of virus DNA synthesis in insect cells (KELLY 1982; ERLANDSON et al. 1985).

Another approach for the identification of potential baculovirus origins of DNA replication has utilized a yeast vector system to detect autonomous replicating sequences (ARS) in *Saccharomyces cerevisiae* (STRUHL et al. 1979). Baculovirus restriction fragments were cloned randomly into YIp5, and putative ARS were identified within the AcMNPV *Eco*RI K and *Eco*RI F fragments (HOOFT VAN IDDEKINGE et al., to be published). We have also identified yeast plasmid chimeras containing the AcMNPV *Eco*RI K fragment, which are capable of transforming *S. cerevisiae* (R. LO and E.B. CARSTENS, unpublished results). Thus far, there has been no direct demonstration that these ARS-containing regions carry natural origins of DNA replication used by the virus in permissive cells.

Aberrations in baculovirus DNA replication in cell culture have also been noted. It has been known for some time that phenotypic changes occur in the production of occlusion bodies following continued passage in cell culture (MACKINNON et al. 1974; POTTER et al. 1976; FRASER and HINK 1982). The change in phenotype from many polyhedra (MP) to few polyhedra (FP) per infected cell nuclei has, in some cases, been associated with genomic alterations (CROZIER and QUOIT 1981; MILLER and MILLER 1982; FRASER et al. 1983; CARSTENS et al. 1986). These have resulted from the insertion of host cell DNA sequences (MILLER and MILLER 1982; FRASER et al. 1983) into the viral genome or the insertion of repeated sequences of viral DNA (BURAND and SUMMERS 1982). A similar process may occur during replication of baculovirus in insects. In the laboratory the selection of FP variants is convenient since it is based on the observation of numbers of OB/per infected cell. FP variants are selected against in insects since the synthesis of fewer OB effectively lowers virulence. Isolates of baculovirus from insects consist of a mixture of genotypic variants (MILLER and DAWES 1978; SMITH and SUMMERS 1978; TJIA et al. 1979) and it is likely that other genomic variants occur during replication of baculovirus, for example those in which some vDNA sequences are reiterated. If these rearrangements result in a lower virulence for the virus, they would be selected against in an insect population.

Two other mutants have been described which demonstrate aberrations in DNA replication. Ts8 synthesizes DNA which contains copies of a 1.35-kbp sequence of viral DNA tandemly repeated in the 90-map-unit region of its genome (ERLANDSON et al. 1984). Variants which carry the insertions apparently are not viable since plaque purification does not result in the isolation of infectious virions missing the original non-repeated DNA sequence. The *Hind*III P fragment, including the 1.35-kb insert, has recently been sequenced (J. KUZIO, P. FAULKNER, C. CURRY, and E.B. CARSTENS, unpublished results) and an open reading frame potentially coding for a 74K protein has been identified. The reiteration in ts8 includes sequences completely within this open reading frame suggesting the presence of an essential gene in this location in the viral genome.

The morphology mutant M5 has also been shown to have an aberration in DNA replication. In this case, two size classes of viral DNA are synthesized, one of 100% and another of 58% of the normal genome length (CARSTENS 1982). The smaller DNA molecule results from the specific deletion (or lack of synthesis) of sequences between 2.6 and 47 map units. M5 carries two almost identical inserts of 290 bases of DNA at these two locations (CARSTENS 1982; E.B. CARSTENS, unpublished results). However, it is not known how the presence of these inserts affects the replication of the M5 genome.

6 Transcription in Baculoviruses

Most of published baculovirus transcription data is based on studies of AcMNPV expression. Biochemical analysis of AcMNPV mRNAs has revealed a cap structure on the 5′ termini, indicating a similarity to other eukaryotic mRNAs (JUN-CHUANG and WEAVER 1982). It is assumed that viral mRNAs are polyadenylated since both total cytoplasmic and poly (A)+ RNA give identical patterns in northern blot analyses (ERLANDSON et al. 1985) and give similar polypeptides in in vitro translation assays (VLAK et al. 1981). AcMNPV-infected cells express a unique RNA polymerase activity which is distinct from the host cell RNA polymerase II (GRULA et al. 1981). The presence of an alpha-amanitin-resistant RNA polymerase activity in infected cells suggests that the virus either codes for its own polymerase or it uses an enzyme other than the host RNA polymerase II for the majority of its transcription (VLAK et al. 1981). It appears that initially virus transcription must utilize host polymerases since purified virus DNA can initiate infections (CARSTENS et al. 1980; BURAND et al. 1980). Later in infection it is possible that virus-specific proteins modify the host polymerase, enabling virus promoters to be recognized (GUARINO and SUMMERS 1986). All the late baculovirus promoters that have been sequenced reveal similar upstream regulatory sequences such as TATA boxes and CAAT sequences (HOOFT VAN IDDEKINGE et al. 1983; KUZIO et al. 1984). These sequences have been associated with RNA polymerase II transcribed genes.

The identification and mapping of viral-encoded mRNAs has been achieved by various techniques including: hybridization selection of mRNA using purified vDNA (SMITH et al. 1982; ROHEL et al. 1984) or cloned restriction fragments of vDNA (ESCHE et al. 1982; SMITH et al. 1983; LUBBERT and DOERFLER 1984a); purification of radiolabeled intracellular RNA and subsequent hybridization to vDNA (ERLANDSON and CARSTENS 1983); northern blot analysis of RNA using cloned fragments of vDNA as probes (ROHEL and FAULKNER 1984; ERLANDSON et al. 1985); and the use of cDNA clones as probes for specific mRNAs (ADANG and MILLER 1982). By following the synthesis of virus-specific RNA at intervals postinfection, a complex pattern of virus transcription has emerged (ROHEL and FAULKNER 1984; ERLANDSON et al. 1985). The overall pattern of virus transcription has been studied by hybridizing ^{32}P-labeled RNA (extracted at time points after infection) with vDNA using a dot-blot procedure (ERLANDSON and CARSTENS 1983). The results indicate that there are two major phases of transcription. The first, which precedes the initiation of vDNA replication, begins within 1 h of infection and continues for about 6 h. The early period encompasses several distinct stages of transcription involving changes in either the recognition of promoter regions or polyadenylation sites of transcripts with the same major body of RNA (LUBBERT and DOERFLER 1984b; FRIESEN and MILLER 1985). The early phase is followed by a dramatic quantitative increase in the level of transcription. In this late stage, transcription appears to continue until the infected cells lyse (ROHEL and FAULKNER 1984).

Two approaches have been used to delineate the location of transcriptionally active regions on the physical map during the replication cycle of AcMNPV. In the first, both total cytoplasmic and poly (A)+RNA, labeled in cell cultures and harvested at intervals postinfection, were hybridized to Southern blots of cloned fragments of AcMNPV DNA (VLAK and VAN DER KROL 1982; ERLANDSON and CARSTENS 1983). Eight specific regions were identified which synthesized major virus transcripts as early as 2.5 h postinfection. There was no apparent clustering of early regions, since the early transcripts hybridized to almost all loci on the genome (ERLANDSON and CARSTENS 1983). However, it was possible to distinguish regions of early transcription based on the relative intensities of hybridization to the DNA fragments, and to estimate the level of transcription. There was an increase in the level of transcription from other regions of the genome which coincided with the initiation of vDNA synthesis at 6 h postinfection, suggesting that expression of early transcripts was necessary before additional regions of the genome could be transcribed. A requirement for early transcription is also suggested by the analysis of virus transcription with a temperature-sensitive mutant, ts8, defective in the synthesis of DNA at the nonpermissive temperature (ERLANDSON et al. 1985). Transcripts normally seen late after infection were absent in infected cells at the restricted temperature. Some overlapping early transcripts of increasing size which hybridized with AcMNPV *Eco*RI J and *Eco*RI L fragments were also detected. This suggests that the defect in ts8 probably does not affect the transcription patterns observed in the alpha and beta phases of polypeptide synthesis (KELLY and LESCOTT 1981; see Sect. 7).

A second approach to study the overall transcription pattern of AcMNPV has used northern blot analysis of cytoplasmic or poly (A)+RNA harvested at increasing times postinfection (ROHEL and FAULKNER 1984; ERLANDSON et al. 1985). The blots were probed with labeled cloned vDNA restriction fragments which taken together included the entire AcMNPV genome. As can be expected from such a large genome, the analysis revealed a large number of sizes and map locations of transcripts. At least 84 different virus-specific transcripts were mapped on the AcMNPV genome in this way (ROHEL and FAULKNER 1984, ERLANDSON et al. 1985). It is not yet established whether the homologous repeat regions of the genome (COCHRAN and FAULKNER 1983) are transcribed. A deletion of sequence from one such region (HR-5) did not affect growth of the virus in cell culture (J. KUZIO, personal communication). The homologous repeats may be important in the regulation of transcription, since in at least one instance (HR-5), transcription terminates at the beginning of the DNA sequence (LUBBERT and DOERFLER 1984b; FRIESEN and MILLER 1985; RANKIN et al. 1986).

An interesting aspect of recent studies on transcription has been the discovery of sets of RNAs which appear to exceed the coding capacity of the genomic segment to which they map. For several regions of the AcMNPV genome, it has been established that nested sets of overlapping transcripts are synthesized with either common 5′ initiation sites or coterminal 3′ polyadenylation sites (LUBBERT and DOERFLER 1984b; FRIESEN and MILLER 1985; RANKIN et al. 1986). The presence of common transcripts with increasing length correlates with increasing times after infection and suggests that the longer transcripts may be involved, at least indirectly, in gene regulation (FRIESEN and MILLER 1985). For example, transcripts with common 3′ ends but different 5′ termini may result from the recognition of transcription initiation sites farther upstream with increasing time postinfection, and coincidentally inhibit recognition of downstream initiation sites. This mechanism could be effective in producing overlapping transcripts, which would code for different polypeptides from a common DNA region. It has yet to be shown that overlapping transcripts code for different proteins, or that only transcripts of a certain length from one particular overlapping region are translationally active.

Overlapping transcripts with common 5′ ends have also been described for the two major late gene products: the transcripts for polyhedrin and for *p10* (FRIESEN and MILLER 1985, RANKIN et al. 1986). In these cases, longer transcripts are synthesized as the infection proceeds, suggesting that there may be some alteration in the recognition of poly[A] addition signals. There seems to be no difference in either the quantity or quality of the gene product synthesized by these longer transcripts. In the case of the *p10* gene, no other open reading frame in the same polarity as *p10* has been found in 2.1 kb of sequence downstream (KUZIO et al. 1984; J. KUZIO, P. FAULKNER, C. CURRY and E.B. CARSTENS, unpublished results). Thus, the longer transcripts do not appear to synthesize a different polypeptide than the shortest transcript. It is possible that the longer transcript may be involved in some other regulatory function, for example the repression of a downstream gene.

Another unique feature of AcMNPV transcription is the apparent absence of RNA splicing (LUBBERT and DOERFLER 1984a). Because the mRNAs are very similar to other eukaryotic mRNAs in that they possess poly[A] tails and are capped at their 5′ ends, the lack of splicing is intriguing. It is possible that the unusual pattern of overlapping transcription may be an alternative mechanism of gene regulation evolved by the baculoviruses.

7 Protein Synthesis in Infected Cells

AcMNPV infects susceptible insect cells in culture and induces a series of events that lead to the production by budding at the cell membrane of mature infectious nonoccluded virus (NOV) by 12 h postinfection (KNUDSON and TINSLEY 1974; VOLKMAN et al. 1976; CARSTENS et al. 1979; TJIA et al. 1979). The replication cycle of the viruses continues for several days. During this late period enveloped nucleocapsids form in the nucleus and become encapsulated within the OB. Several groups of investigators have studied the temporal regulation of protein synthesis in cells infected with baculoviruses (CARSTENS et al. 1979; DOBOS and COCHRAN 1980; WOOD 1980; MARUNIAK and SUMMERS 1981; VLAK and VAN DER KROL 1982; MCCLINTOCK et al. 1986). A sequential and regulated expression of the viral genome throughout the replication cycle is seen and translation products are detected from as early as 3 h to at least 60 h postinfection. About 40 different polypeptides, specific for infected cells, have been identified by one-dimensional SDS-PAGE of [^{35}S]methionine-labeled cell extracts. Since the resolving power of one-dimensional gels is limited, there may well be over 100 virus-coded proteins synthesized in infected cells. In other studies of virus structural proteins 80–95 polypeptides were detected using two-dimensional gel electrophoresis (SINGH et al. 1983; SINGH et al. 1985). Since many regulatory gene products and other nonstructural proteins will be synthesized in quantities too low to detect by these methods, other approaches will be necessary to identify them in infected cells. In a recent study recombinant plasmids carrying the chloramphenicol acetyl transferase (CAT) gene were linked to genomic fragments to monitor the promoter activity of an AcMNPV regulatory gene (GUARINO and SUMMERS 1986). It can be anticipated that this approach will allow the identity and activity of other regulatory elements to be studied.

Up to five virus-induced polypeptides have been detected in cell extracts by pulse labeling infected cells at very early times after infection (CARSTENS et al. 1979; KELLY and LESCOTT 1983; WOOD 1980). The polypeptide profile does not change until 6–8 h postinfection when many more virus-induced polypeptides appear. During this later phase of the virus replication cycle, quantitative regulation of protein synthesis occurs since many proteins are synthesized in varying amounts and for various lengths of time. Because of these temporal changes, the infectious cycle has been divided into three, four, or five arbitrary stages (CARSTENS et al. 1979; DOBOS and COCHRAN 1980; WOOD 1980; MARUNIAK and SUMMERS 1981; KELLY and LESCOTT 1981). The alpha phase (0–3 h postinfection includes the synthesis of virus-specific polypeptides which do not require prior viral protein synthesis. Presumably, genes expressed during this phase are entirely dependent upon the host cell transcription and translation systems for their production. It has been suggested that these alpha proteins are the ones synthesized soon after the uncoated genome is transcribed in the nucleus, or when experimental inhibitors of protein synthesis have been removed (KELLY and LESCOTT 1983).

Synthesis of proteins required for initiation of DNA replication is believed to occur in the beta phase at 3–6 h postinfection. It is in this phase that the defect in the mutant ts8 is considered to occur (GORDON and CARSTENS 1984). A normal complement of early protein is seen in ts8-infected cells at the non-permissive temperature, but these are synthesized continuously in the absence of DNA replication. Many alpha and beta proteins can be detected at late times postinfection; thus they may be synthesized continuously once transcription is initiated.

Following the onset of DNA replication, another group of polypeptides is synthesized in the gamma phase (6–10 h postinfection). These proteins are considered to be predominantly structural proteins required for the assembly of budded virions destined to migrate into the extracellular media. The delta phase follows at later than 10 h postinfection. It is characterized by the synthesis of proteins necessary for the processing of OV into OB. It is during this stage that the two most abundant baculovirus proteins are synthesized, polyhedrin and *p10* (ROHEL et al. 1983; SMITH et al. 1983). The stages outlined here have not always been observed in studies of the time course of protein synthesis probably due to the variety of experimental conditions used such as differences in MOI, the host cells and virus strains, as well as the continued expression of many gene products after their induction. Studies with ts mutants and analysis of RNA transcription patterns confirm the time frame for the various stages of AcMNPV polypeptide synthesis. The use of sensitive assay systems such as the "CAT" gene expression system should improve the resolution of the various stages in the cascade of virus gene expression. This in turn will allow the functions of the virus coded proteins to be determined.

8 Conclusions

The quest for alternative pesticides has stimulated much of the recent research into the molecular biology of baculoviruses and many of the tools of recombinant DNA technology are now being applied to their study. A better understanding of the biological machanisms involved in the control of expression of the genome will enable rational decisions to be made in integrated pest management programs. It is anticipated that novel baculovirus strains will be engineered having altered insect host ranges and virulence for target pests. In addition the new strains will incorporate genes which when expressed will affect the physiology of the pest species. Another focus of recent research has been the construction of baculovirus expression vectors of great potential for the manufacture of diverse gene products in an insect cell background. There is still a great deal to learn about the genome and its expression and continued in depth basic investigations are justified since future innovative applications of benefit to mankind seem assured.

Acknowledgments. We thank Mark A. Cochran and Lois K. Miller for making manuscripts of their reviews available before publication. We also acknowledge the financial support of the Medical Research Council of Canada.

References

Acqua C (1918) Richeche sulla malattia del Rend. 1st bacol Scuola super agr Protici 3:243

Adams J, Goodwin RH, Wilcox TA (1977) Electron microscope investigation on invasion and replication of insect baculoviruses *in vivo* and *in vitro*. Biol Cellulaire 28:261–268

Adang MJ, Miller LK (1982) Molecular cloning of DNA complementary to mRNA of the baculovirus *Autographa californica* nuclear polyhedrosis virus: lacation and gene products of RNA transcripts found late in infection. J Virol 44:782–793

Arif B, Brown K (1975) Purification and properties of a nuclear polyhedrosis virus form *Choristoneura fumiferana*. CJ Microbiol 21:1224–1231

Arif B, Doerfler W (1984) "Identification and localization of reiterated sequences in the *Choristoneura fumiferana* MNPV genome. EMBO J 3:525–529

Arif BM, Kuzio J, Faulkner P, Doerfler W (1984) The genome of *Choristoneura fumiferana* nuclear polyhedrosis virus: molecular cloning and mapping of the *Eco*RI, *Bam*HI, *Sma*I, *Xba*I, and *Bgl*I restriction sites. Virus Res 1:605–614

Arif BM, Tjia ST, Doerfler W (1985) DNA homologies between the genomes of *Choristoneura fumiferana* and *Autographa californica* nuclear polyhedrosis viruses. Virus Res 2:85–94

Bassemir U, Miltenberger HG, David P (1983) Morphogenesis of nuclear polyhedrosis virus from *Autographa californica* in a cell line from *Mamestra brassicae* (cabbage moth). Further aspects on baculovirus assembly. Cell Tissue Res 228:587–595

Beaton CD, Filshie BK (1976) Comparative ultrastructural studies of insect granulosis and nuclear polyhedrosis viruses. J Gen Virol 31:151–161

Bergold GH (1947) Die Isolierung des Polyeder-virus und die Natur der Polyeder. Z Naturforsch 2b:122–143

Bergold GH (1953) Insect viruses. Adv Virus Res 1:91–139

Bilimoria SL (1983) Genomic divergence among single-nucleocapsid nuclear polyhedrosis viruses of plusiine hosts. Virology 127:15–23

Blinov VM, Gutorov VV, Holodilov GG, Iljchev AA, Karginov VA, Mikrjukov NN, Mordvinov VA, Nikonov IV, Petrov NA, Urmanov IH, Vasilenko SK (1984) Nucleotide sequence of the *Galleria mellonella* nuclear polyhedrosis virus origin of replication. FEBS 167:254–258

Brown M, Crawford AM, Faulkner P (1979) Genetic analysis of a baculovirus *Autographa californica* nuclear polyhedrosis virus. I. Isolation of temperature-sensitive mutants and assortment into complementation groups. J Virol 31:190–198

Brown SE, Maruniak JE, Knudson DL (1984) Physical map of SeMNPV baculovirus. Virology 136:235–240

Brown SE, Maruniak JE, Knudson DL (1985) Baculovirus MNPV genomic variants: Characterization of *Spodoptera exempta* MNPV DNA and comparison with other *Autographa californica* MNPV DNA. J gen Virol 66:2431–2441

Burand JP, Summers MD (1982) Alterations of *Autographa* californica nuclear polyhedrosis virus upon serial passage in cell culture. Virology 119:223–229

Burand J, Summers M, Smith G (1980) Transfection with baculovirus DNA. Virology 101:286–290

Carstens EB (1982) Mapping the mutation site of an *Autographa californica* nuclear polyhedrosis virus polyhedron morphology mutant. J Virol 43:809–818

Carstens EB, Tjia ST, Doerfler W (1979) Infection of *Spodoptera frugiperda* cells with *Autographa californica* nuclear polyhedrosis virus. I. Synthesis of intracellular proteins after virus infection. Virology 99:386–398

Carstens EB, Tjia ST, Doerfler W (1980) Infectious DNA from *Autographa californica* nuclear polyhedrosis virus. Virology 101:311–314

Carstens EB, Krebs A, Gallerneault CE (1986) Identification of an amino acid essential to the normal assembly of *Autographa californica* nuclear polyhedrosis virus polyhedra. J Virol 58:684–688

Chung K, Brown M, Faulkner P (1980) Studies on the morphogenesis of polyhedral inclusion bodies of a baculovirus *Autographa californica* NPV. J Gen Virol 46:335–347
Cochran MA, Faulkner P (1983) Location of homologous DNA sequences interspersed at five regions of the baculovirus AcMNPV genome. J Virol 45:961–970
Cochran MA, Carstens EB, Eaton BT, Faulkner P (1982) Molecular cloning and physical mapping of restriction endonuclease fragments of *Autographa californica* nuclear polyhedrosis virus DNA. J Virol 41:940–946
Cochran MA, Brown SE, Knudson DL (to be published) Organization and expression of the baculovirus genome. In: Granados RR (ed) The baculoviruses. CRC Press
Crawford AM, Sheehan C (1985) Replication of *Oryctes* baculovirus in cell culture: viral morphogenesis, infectivity and protein synthesis. J Gen Virol 66:529–539
Crawford AM, Ashbridge K, Sheehan C, Faulkner P (1985) A physical map of the *Oryctes* baculovirus genome. J Gen Virol 66:2649–2658
Crozier G, Quoit JM (1981) Obtention and analysis of two genetic recombinants of baculoviruses of Lepidoptera, *Autographa californica* Speyer and *Galleria mellonella* L. Annales de Virologie 132:3–18
Dobos P, Cochran MA (1980) Protein synthesis in cells infected by *Autographa californica* nuclear polyhedrosis virus (Ac-NPV): the effect of cytosine arabinoside. Virol 103:446–464
Dougherty EM, Vaughn JL, Reichelderfer CF (1975) Characteristics of the non-occluded from of a nuclear polyhedrosis virus. Intervirol 5:109–121
Erlandson MA, Carstens EB (1983) Mapping early transcription products of *Autographa californica* nuclear polyhedrosis virus. Virology 126:398–402
Erlandson MA, Skepasts P, Kuzio J, Carstens EB (1984) Genomic variants of a temperature-sensitive mutant of *Autographa californica* nuclear polyhedrosis virus containing specific reiterations of viral DNA. Virus Res 1 :565–584
Erlandson MA, Gordon J, Carstens EB (1985) Size and map locations of early transcription products on the *Autographa californica* nuclear polyhedrosis virus genome. Virology 142:12–23
Esche H, Lubbert H, Siegmann B, Doerfler W (1982) The translational map of the *Autographa californica* nuclear polyhedrosis virus (AcNPV) genome. EMBO 1:1629–1633
Faulkner P (1981) Baculovirus: In: Davidson EW (ed), Pathogenesis of invertebrate microbial diseases. Allanheld, Osmun and Co Ltd, Totowa, New Jersey, pp 3–33
Fraser MJ, Hink W (1982) The isolation and characterization of the MP and FP variants of *Galleria mellonella* nuclear polyhedrosis virus. Virology 117:366–378
Fraser MJ, Smith GE, Summers MD (1983) Acquisition of host cell DNA sequences by baculoviruses: relationship between host DNA insertions and FP mutants of *Autographa californica* and *Galleria mellonella* nuclear polyhedrosis virus. J Virol 47:287–300
Fraser MJ, Brusca JS, Smith GE, Summers MD (1985) Transposon mediated mutagenesis of a baculovirus. Virology 145:356–361
Friesen P, Miller L (1985) Temporal regulation of baculovirus RNA: overlapping early and late transcripts. J Virol 54:392–400
Gordon JD, Carstens EB (1984) Phenotypic characterization and physical mapping of a temperature-sensitive mutant of *Autographa californica* nuclear polyhedrosis virus defective in DNA synthesis. Virology 138:69–81
Granados RR (1980) Infectivity and mode of action of baculoviruses. Biotechnology and Bioengineering 22:1377–1405
Grula MA, Buller PL, Weaver RF (1981) Alpha-Amanitin resistant viral RNA synthesis in nuclei isolated from nuclear polyhedrosis virus infected *Heliothis zea* larvae and *Spodoptera frugiperda* cells. J Virol 38:916–921
Guarino L, Summers M (1986) Functional mapping of a *trans*-activating gene required for expression of a delayed-early gene. J Virol 57:563–571
Harrap KA (1972) The structure of nuclear polyhedrosis viruses. II. The virus particle. Virology 50:124–132
Harrap KA, Payne CC (1979) The structural properties and identification of insect viruses. Adv Vir Res 25:273–355
Harvey JP, Volkman LE (1983) Biochemical and biological variation of *Cydia pomonella* (codling moth) granulosis virus. Virology 124:21–34
Hooft van Iddekinge BJL, Smith GE, Summers MD (1983) Nucleotide sequence of the polyhedrin gene of *Autographa californica* nuclear polyhedrosis virus. Virology 131:561–565

Jewell J, Miller L (1980) DNA sequence homology relationships among six lepidopteran nuclear polyhedrosis viruses. J Gen Virol 48:161–175

Jun-Chuang Q, Weaver RF (1982) Capping of viral RNA in cultured *Spodoptera frugiperda* cells infected with *Autographa californica* nuclear polyhedrosis virus. J Virol 43:234–240

Keddie B, Volkman L (1985) Infectivity difference between the two phenotypes of *Autographa californica* nuclear polyhedrosis virus: importance of the 64K envelope glycoprotein. J Gen Virol 66:1195–1200

Kelly DC (1982) Baculovirus replication. J Gen Virol 63:1–13

Kelly DC (1985) The structure and physical characteristics of baculoviruses. In: Maramorosch K, Sherman KE (eds) Viral insecticides for biological control. Academic, New York, pp 469–488

Kelly DC, Lescott T (1976) The effects of inhibitors of nucleic acid and protein synthesis on the replication of *Spodoptera frugiperda* nuclear polyhedrosis virus in cultured *Spodoptera frugiperda* cells. Intervirol 7:360–365

Kelly DC, Lescott T (1981) Baculovirus replication: protein synthesis in *Spondoptera frugiperda* cells infected with *Trichoplusia ni* nuclear polyhedrosis virus. Microbiologica 4:35–57

Kelly DC, Lescott T (1983) Baculovirus replication: phosphorylation of polypeptides synthesized in *Trichoplusia ni* nuclear polyhedrosis virus-infected cells. J Gen Virol 65:1183–1191

Kelly DC, Wang X (1981) The infectivity of nuclear polyhedrosis virus DNA. Annales de Virologie 132E:247–259

Kelly DC, Brown DA, Ayers MD, Allen CJ, Walker IO (1982) Properties of the major nucleocapsid protein of *Heliothis zea* nucleocapsid nuclear polyhedrosis virus. J Gen Virol 62:313–322

Knell JD, Summers MD (1981) Investigation of genetic heterogeneity in wild isolates of *Spodoptera frugiperda* nuclear polyhedrosis virus by restriction endonuclease analysis of plaque-purified variants. Virology 112:190–197

Knell JD, Summers MD (1984) A physical map for the *Heliothis zea* SNPV genome. J Gen Virol 65:445–450

Knudson DL, Buckley SM (1977) Invertebrate cell culture: methods for the study of invertebrate associated animal viruses. In: Maramorosch K, Kropowski H (eds) Methods in virology, vol 6. Academic, New York, pp 323–391

Knudson DL, Tinsley TW (1974) Replicating nuclear polyhedrosis virus in a continuous cell culture of *Spodoptera frugiperda*: purification, assay of infectivity and growth characteristics of the virus. J Virol 14:934–944

Kuzio J, Rohel DZ, Curry CJ, Krebs A, Carstens EB, Faulkner P (1984) Nucleotide sequence of the *p10* polypeptide gene of *Autographa californica* nuclear polyhedrosis virus. Virology 139:414–418

Lee HH, Miller LK (1978) Isolation of genotypic variants of *Autographa californica* nuclear polyhedrosis virus. J Virol 27:754–767

Leisy D, Rohrmann G, Beaudreau G (1984) Conservation of genome organization in two multicapsid nuclear polyhedrosis viruses. J Gen Virol 52:699–702

Lubbert H, Doerfler W (1984a) Mapping of early and late transcripts encoded by the *Autographa californica* nuclear polyhedrosis virus genome: is viral RNA spliced. J Virol 50:497–506

Lubbert H, Doerfler W (1984b) Transcription of overlapping sets of RNAs from the genome of *Autographa californica* nuclear polyhedrosis virus: a novel method for mapping RNAs. J Virol 52:255–265

Lubbert H, Kruczekg S, Tjia S, Doerfler W (1981) The cloned *Eco*RI fragments of *Autographa californica* nuclear polyhedrosis virus DNA. Gene 16:343–345

MacKinnon EA, Henderson JF, Stoltz DB, Faulkner P (1974) Morphogenesis of nuclear polyhedrosis virus under conditions of prolonged passage *in vitro*. J Ultrastruct Res 49:419–435

Maruniak JE, Summers MD (1981) *Autographa californica* nuclear polyhedrosis virus phosphoproteins and synthesis of intracellular proteins after virus infection. Virology 109:25–34

Maruniak JE, Brown SE, Knudson DL (1984) Physical maps of SfMNPV baculovirus DNA and its genomic variants. Virology 136:221–234

Matthews REF (1982) Classification and nomenclature of viruses. Fourth report of the international committee on taxonomy of viruses. Intervirol 17:1–199

McClintock J, Dougherty E, Weiner R (1986) Semipermissive replication of a nuclear polyhedrosis virus of *Autographa californica* in a gypsy moth cell line. J Virol 57:197–204

Miller DW, Miller LK (1982) A virus mutant with an insertion of a copia-like transposable element. Nature 299:562–564

Miller LK (1984) Exploring the genome organization of baculoviruses. In: Methods in virology, Maramorosch K, Koprowski H vol 7. Academic, New York, pp 227–258

Miller LK, Dawes KP (1978) Restriction endonuclease analysis for the identification of baculovirus pesticides. Appl Environ Microbiol 35:411–421

Miller LK, Dawes KP (1979) Physical map of the DNA genome of *Autographa californica* nuclear polyhedrosis virus. J Virol 29:1044–1055

Miller LK, Miller DW, Adang MJ (1982) An insect virus for genetic engineering: developing baculovirus polyhedrin substitution vectors. In: Lurquin PF, Kleinhofs A (eds) Genetic engineering in eukaryotes NATO advanced sciences institute, Series A vol 61. Plenum, New York, pp 89–98

Miller LK, Ling AJ, Bulla LAJ (1983a) Bacterial, viral and fungal insecticides. Science 219:715–721

Miller LK, Trimarchi RE, Browne D, Pennock GD (1983b) A temperature-sensitive mutant of the baculovirus *Autographa californica* nuclear polyhedrosis virus defective in an early function required for further gene expression. Virology 126:376–380

Miltenburger HG, Naser WL, Harvey JP (1984) The cellular substrate: very important requirement for baculovirus *in vitro* replication. Z Naturforsch 39c:993–1002

Naser WL, Miltenburger HG, Harvey JP, Huber J, Huger AM (1984) In vitro repication of the *Cydia pomella* (codling moth) granulosis virus. FEMS Microbiol Letters 24:117–121

Potter KN, Miller LK (1980) Transfection of two invertebrate cell lines with DNA of *Autographa californica* nuclear polyhedrosis virus. J Invertebr Pathol 36:431–432

Potter KN, Faulkner P, MacKinnon EA (1976) Strain selection during serial passage of *Trichoplusia ni* nuclear polyhedrosis virus. J Virol 18:1040–1050

Priston RAJ (1985) Production of insect pathogenic viruses in cell culture. Animal Cell Biotechnology 2:71–83

Prowazek S von (1912) Untersuchungen über die Gelbsucht der Seidenraupen. Zentr Bakt Parasitenk Abt I Orig 67:268

Ralston AL, Huang YS, Kawanishi CY (1981) Cell culture studies with the IMG-Hz 1 nonoccluded virus. Virology 115:33–44

Rankin C, Ladin B, Weaver R (1986) Physical mapping of temporally regulated, overlapping transcripts in the region of the 10K protein gene in *Autographa californica* nuclear polyhedrosis virus. J Virol 57:18–27

Rohel DZ, Faulkner P (1984) Time course analysis and mapping of *Autographa californica* NPV transcripts. J Virol 50:739–747

Rohel DZ, Cochran MA, Faulkner P (1983) Characterization of two abundant mRNAs of *Autographa californica* nuclear polyhedrosis virus present late in infection. Virology 124:357–368

Rohel DZ, Cochran MA, Faulkner P (1984) Mapping of late transcripts of *Autographa californica* nuclear polyhedrosis virus by "criss-cross" DNA-RNA hybridization. J Gen Virol 65:809–813

Rohrmann GF, Pearson MN, Bailey TJ, Becker RR, Beaudreau GS (1981) N-terminal polyhedrin sequences and occluded baculovirus evolution. J Molec Evol 17:329–333

Singh SP, Gudauskas RT, Harper JD (1983) High resolution two-dimensional gel electrophoresis of structural proteins of baculoviruses of *Autographa californica* and *Porthetria (Lymantria)*. Virology 125:370–380

Singh S, Gudauskas R, Harper J, Edwards J (1985) Two-dimensional gel electrophoresis of basic proteins of *Autographa californica* nuclear polyhedrosis virus. J Invertebr Pathol 45:249–251

Smith GE, Summers MD (1978) Analysis of baculovirus genomes with restriction endonucleases. Virology 89:517–527

Smith GE, Summers MD (1979) Restriction maps of five *Autographa californica* MNPV variants, *Trichoplusia ni* MNPV, and *Galleria mellonella* MNPV DNAs with endonucleases *Sma*I, *Kpn*I, *Bam*HI, *Sac*I, *Xho*I, and *Eco*RI. J Virol 30:828–838

Smith GE, Summers MD (1982) DNA homology among subgroup A, B, and C beculoviruses. Virology 123:393–406

Smith GE, Vlak JM, Summers MD (1982) *In vitro* translation of *Autographa californica* nuclear polyhedrosis virus early and late mRNA. J Virol 44:199–208

Smith GE, Vlak JM, Summers MD (1983) Physical analysis of *Autographa californica* nuclear polyhedrosis virus transcripts for polyhedrin and 10,000 molecular-weight protein. J Virol 45:215–225

Stiles B, Wood HA (1983) A study of the glycoproteins of *Autographa californica* nuclear polyhedrosis virus (AcMNPV). Virology 131:230–241

Stoltz DB, Pavan C, DaCuna AB (1973) Nuclear polyhedrosis virus: possible example of *de novo* intranuclear membrane morphogenesis. J Gen Virol 19:145–150

Struhl K, Stinchcomb DT, Scherer S, Davis RW (1979) High-frequency transformation of yeast: autonomous replication of hybrid DNA molecules. PNAS 76:1035–1039
Summers MD, Smith GE (1978) Baculovirus structural polypeptides. Virology 84:390–402
Summers MD, Smith GE, Knell JD, Burand JP (1980) Physical maps of *Autographa californica* and *Rachiplusia ou* nuclear polyhedrosis virus recombinants. J Virol 34:693–703
Tinsley TW, Harrap KA (1978) Viruses of invertebrates. In: Fraenkel-Conrat H, Wagner RR (eds) Comprehensive virology, vol 12, pp 1–101 Plenum, New York
Tjia ST, Carstens EB, Doerfler W (1979) Infection of *Spodoptera frugiperda* cells with *Autographa californica* nuclear polyhedrosis virus. II. The viral DNA and the kinetics of its replication. Virology 99:391–409
Tweeten KA, Bulla LA, Consigli RA (1980a) Restriction analysis of the genomes of *Plodia interpunctella* and *Pieris rapae* granulosis viruses. Virology 104:514–519
Tweeten KA, Bulla LA, Consigli RA (1980b) Characterization of an extremely basic protein derived from granulosis virus nucleocapsids. J Virol 33:866–876
Tweeten KA, Bulla LA, Consigli RA (1981) Applied and molecular aspects of insect granulosis viruses. Microbiol Rev 45:379–408
Vaughn JL, Dougherty EM (1985) The replication of baculoviruses. In: Maramorosch K, Sherman KE (eds) Viral insecticides for biological control. Academic, New York, pp 569–633
Vlak JM (1980) Mapping of *Bam*HI and *Sma*I restriction sites on the genome of the nuclear polyhedrosis virus of the alfalfa looper *Autographa californica.* J Invertebr Pathol 36:409–414
Vlak JM, Krol S van der (1982) Transcription of the *Autographa californica* nuclear polyhedrosis virus genome: location of late cytoplasmic RNA. Virology 123:222–228
Vlak JM, Smith GE (1982) Orientation of the genome of *Autographa californica* nuclear polyhedrosis virus: a proposal. J Virol 41:1118–1121
Vlak JM, Smith GE, Summers MD (1981) Hybridization selection and *in vitro* translation of *Autographa californica* nuclear polyhedrosis virus mRNA. J Virol 40:762–771
Volkman LE (1983) Occluded and budded *Autographa californica* nuclear polyhedrosis virus: immunological relatedness of structural proteins. J Virol 46:221–229
Volkman LE, Goldsmith PA (1985) Mechanism of neutralization of budded *Autographa californica* nuclear polyhedrosis virus by a monoclonal antibody: inhibition of entry by adsorptive endocytosis. Virology 143:185–195
Volkman LE, Summers MD, Hsieh C-H (1976) Occluded and nonoccluded nuclear polyhedrosis virus grown in *Trichoplusia ni*: comparative neutralization, comparative infectivity, and *in vitro* growth studies. J Virol 19:820–832
Wiegers FP, Vlak JM (1984) Physical map of a *Mamestra brassicae* nuclear polyhedrosis virus variant isolated from *Spodoptera exigua.* J Gen Virol 65:2011–2019
Wilson ME, Consigli RA (1985a) Functions of a protein kinase activity associated with purified capsids of the granulosis virus infecting *Plodia interpunctella.* Virology 143:526–535
Wilson ME, Consigli RA (1985b) Characterization of a protein kinase activity associated with purified capsids of the granulosis virus infecting *Plodia interpunctella.* Virology 143:516–525
Wood HA (1980) *Autographa californica* nuclear polyhedrosis virus-induced proteins in tissue culture. Virology 102:21–27

The Structure of the Viral Genome

B.M. ARIF

1 Introduction

The genomes of baculoviruses consist of double-stranded, circular, supercoiled DNA ranging in size from 90 to 200 kb as determined by electron microscopy, sedimentation through sucrose gradients, and digestion with restriction endonuclease enzymes (RENs) followed by electrophoresis. The genome comprises approximately 13% of the virus particle by weight. The sites of many RENs on the genomes of several baculoviruses have been mapped, which gave an insight into the structural arrangement of these large and complex DNAs. The genomes were found to contain reiterated sequences, the function of which is still to be elucidated. Host DNA sequences can become incorporated in the genome and result in the formation of defective particles. Several investigators studied the DNA:DNA homology among baculoviruses and showed that representatives of subgroups A, B, and C have homologous DNA sequences indicating that they evolved from a common ancestor. The base composition, buoyant density data, and T_m analyses will not be reviewed here. For these, the reader is referred to an earlier review by HARRAP and PAYNE (1979).

Aspects of the structure of the genomes of baculoviruses will be reviewed in more detail in this section and this will, hopefully, complement rather than reiterate parts of other chapters.

2 Physical Maps

Restriction endonuclease enzymes have served to distinguish isolates of baculoviruses and to generate physical maps of the genomes as a prelude for more

Forest Pest Management Institute, Canadian Forestry Service, PO Box 490, Sault Ste. Marie, Ontario, P6A 5M7, Canada

Current Topics in Microbiology and Immunology, Vol. 131

detailed studies on the transcription, variation, and replication. An example is the physical map of *Choristoneura fumiferana* MNPV (CfMNPV) is shown in Fig. 1 (ARIF et al. 1984). Maps of other baculovirus genomes are shown in other chapters of this book.

Baculoviruses are known to produce naturally occurring variants whose genomes have been shown to contain differences in the restriction endonuclease patterns (SMITH and SUMMERS 1978; LEE and MILLER 1978; MARUNIAK et al. 1984; BROWN et al. 1984). The genomic changes that occur have been located on the physical maps of several variants (MILLER and DAWES 1979; SMITH and SUMMERS 1979; BROWN et al. 1984; MARUNIAK et al. 1984; BROWN et al. 1985; CROOK et al. 1985). Physical maps were also useful in identifying naturally occurring recombinants. Two such viruses AR7 and AR8 were isolated, plaque purified, and analyzed by RENs. They appeared to be recombinants between *Autographa californica* MNPV (AcMNPV) and *Rachiplusia ou* MNPV (RoMNPV) since their genomes contained *Eco*RI, *Bam*HI, and *Xba*I fragments with mobilities which are (a) the same as the RoMNPV DNA fragments, (b) the same as the AcMNPV DNA fragments, or (c) different from RoMNPV and AcMNPV fragments (SMITH and SUMMERS 1980). The recombinants also contained structural polypeptides from both parental types. Peptide mapping of the polyhedrins showed both recombinants were occluded in AcMNPV polyhedrin.

Recombinants were generated in tissue culture by infecting TN-368 cells with nondefective plaque-purified isolates of AcMNPV and RoMNPV (S1 and R1, respectively). Out of 100 plaque isolates 7 were true recombinants. They were isolated and the crossover locations were mapped (SUMMERS et al. 1980). In six out of the seven recombinants, the crossover sites were in the *Eco*RI-Q and -B fragments of RoMNPV and the *Eco*RI-I and -B fragments of AcMNPV genomes. These authors also reported that the crossover events in AR7 and AR8 were in the same regions of the genomes, indicating that the distribution of recombination sites does not appear to be random.

3 Reiterated Sequences in the DNA

The presence of reiterated sequences within the baculovirus genome was first shown in four NPVs infecting *Spodoptera* spp. (KELLY 1977). Reassociation kinetics experiments revealed that a small fraction of the genome (less than 3%) reannealed rapidly indicative of repeated sequences. Similarly, the DNA from a nonoccluded baculovirus (Hz-1V) contained a rapidly reannealing fraction that comprised about 13% of the total viral genome (HUANG et al. 1982). A domain of approximately 11% of the total genome consists of 69% G+C residues and the authors mention the possibility that this domain might have caused the observed rapid reannealing.

In AcMNPV, hybridization was observed among the fragments *Hind*III-Q, -L, -F, and -A/B. Clones of *Hind*III-L and -Q were then used to map the reiterated sequences on the AcMNPV genome. Five locations of homologous

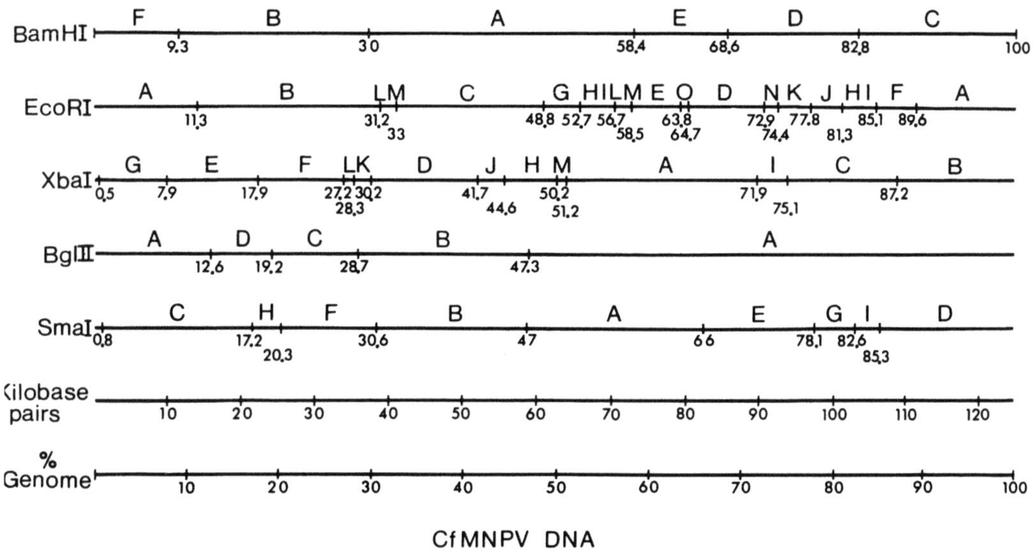

Fig. 1. Physical maps of the CfMNPV genome. The polyhedrin gene is located in the *Bam*HI-F fragment

repeats (hr) were identified at the following junctions: hr_1, *Eco*RI-B-*Eco*RI-I; hr_2, *Eco*R1-A-*Eco*R1-J; hr_3, *Eco*R1-C-*Eco*R1-G; hr_4, *Eco*R1-Q-*Eco*RI-L; and hr_5, *Eco*R1-S-*Eco*R1-X (Cochran and Faulkner 1983). The homologous regions were rich in *Eco*RI sites and contained minifragments ranging in size from 73 to 225 bp. The presence and distribution of the homologous sequences were shown to be conserved among different strains of AcMNPV as were the stoichiometric quantities and the pattern of the minifragments, indicating that the homologous regions could be important in the replication and expression of the AcMNPV genome.

In other series of experiments, AcMNPV was serially passaged in TN-368 cell line for 30 generations and plaques of many polyhedra phenotypes (MP) were isolated and purified. Seven plaque-purified isolates were found to have an altered *Eco*RI pattern and were grouped into three distinct types designated B5, B6, and C6, each type containing multiple copies of additional *Eco*RI fragments. The B5 isolate had three additional 1.97-kb fragments which comigrated with the *Eco*RI-P fragment. The B6 isolate acquired four additional 1.55-kb fragments. The C6 isolate had additional fragments of three distinct sizes: a 3.46–kb fragment present in a 2 *M* amount, a 1.55-kb fragment present in 4 *M* amount, and a 1.47-kb fragment present in 3 *M* amount. Mapping data showed that the additional *Eco*RI fragments originated from locations on the genome close to where the insertions were mapped (Burand and Summers 1982). Hybridization of the additional *Eco*RI fragments to *Sac*I- or *Eco*RI-restricted AcMNPV indicated that they shared homology to two distinct regions of the genome. When *T. ni* larvae were inoculated with the serial passage isolates and the DNA extracted from occluded virus, it was observed that this DNA had an *Eco*RI pattern similar to that produced in cell culture but each additional band was present in a lower concentration. BURAND and SUMMERS (1982) mentioned that the DNA molecules, which contained amplified sequences probably had a selective advantage in cell culture.

Several other baculoviruses were shown to contain reiterated sequences in their genomes. The DNA of *Orgyia pseudotsugata* MNPV (OpMNPV) contained such sequences but their size has not yet been calculated (LEISY et al. 1984). The genome of CfMNPV also contains reiterated sequences dispersed in four locations (ARIF and DOERFLER 1984; KUZIO and FAULKNER 1984). When nick-translated *Eco*RI-E or *Hind*III-R fragments were hybridized to Southern blots of *Hind*III-restricted CfMNPV DNA, a "ladder" of fragments was seen in the autoradiograms. The lower three steps of the ladder represented the fragments *Hind*III-O, -PQ, and -R. The other steps of the ladder (five fragments) were submolar in amount in that they could not be seen in ethidium bromide-stained gels. Recently, more than 15 ladder fragments were observed (ARIF, unpublished data). The fragment *Hind*-R was located in the fragment *Eco*RI-A. All the other ladder fragments were located in the fragment *Eco*RI-B of the genome. Since the CfMNPV was originally plaque purified, then propagated in larvae, ARIF and DOERFLER (1984) concluded that the submolar fragments represented a newly generated virus population, each being distinguished by additional insertion of the repeated DNA in the *Eco*RI-B part of the genome. Recently, the fragment *Hind*III-R of CfMNPV was sequenced and revealed five repeats of 61 bp in the 5′ half of the fragment (J. SPENCER, personal communication). Two were direct tandom repeats followed by the first 19 bp of the third repeat. The third repeat contained unique insertions of 32 bp and 85 bp. The last two were repeated in tandom and exhibited close homology to the first two repeats.

The function of repeated DNA in baculoviruses is still not clear. COCHRAN and FAULKNER (1983) noted that the DNA of the plaque morphology mutant FP-D described by POTTER and MILLER (1980) is altered near the hr_5 sequence of AcMNPV. Also the serial passage mutants reported by BURAND and SUMMERS (1982) appeared to have been duplication within hr_2 and hr_3. A small insertion present in many AcMNPV variants and in GmMNPV reported by SMITH and SUMMERS (1979) was close to hr_4. Taken together, along with data on *Spodoptera exempta* MNPV (SeMNPV) DNA in the chapter on genetic variants, the evidence points toward functional importance and locational significance of the repeated sequences.

4 Genomic Homologies Among Baculoviruses

The earliest report on homologies among the genomes of baculoviruses was that of KELLY (1977), who studied the reassociation kinetics of four genomes from *Spodoptera* NPVs and showed that the genomes were 15%–75% homologous. Homology among the genomes from subgroup A, B, and C baculoviruses were estimated under hybridization conditions of different stringencies (SMITH and SUMMERS 1982). Under the highest stringency condition (Tm-22° C), the apparent homology ranged from 0.1% to 3%. When the concentration of formamide was reduced to 20% (Tm-44° C), there was a three- to ten fold increase in the apparent homology. Hybridization to specific *Eco*RI fragments of AcMNPV showed that DNAs from subgroups A and B hybridized to *Eco*RI-L and -M. The significance of this is still not yet clear. It was suggested that these regions may contain conserved sequences that have an important function such as the origin(s) of viral DNA replication (SMITH and SUMMERS 1982). DNAs from all subgroups hybridized to the *Eco*RI-I fragment of AcMNPV, which is the fragment that contains the polyhedrin gene. Homology between the *Hind*III-V of AcMNPV, which is within the polyhedrin gene, and baculoviruses from *Spodoptera* and *Heliothis* spp. has also been illustrated (KISLEV 1985). The hybridization patterns of OpMNPV and CfMNPV were very similar, which suggested that these viruses are closely related. The genomes of three NPVs and those of two GVs hybridized to the *Eco*RI-P fragment of AcMNPV. This fragment contains the *p10* gene, known to be under the control of a strong promotor and codes for a 10K protein late in the infectious cycle (KUZIO et al. 1984, KNEBEL et al. 1985). The fact that the genomes from subgroup A, B, and C baculoviruses have homologous DNA sequences is good evidence that they evolved from a common ancestor (SMITH and SUMMERS 1982).

A close relationship between the genomes of AcMNPV and OpMNPV was shown by hybridization under varying conditions of stringency (ROHRMANN et al. 1982). Homology in the order of 13%–25% was detected. OpSNPV, on the other hand, had 20% apparent homology to the AcMNPV genome. The close relationship between OpMNPV and AcMNPV was confirmed when cosmid clones of the OpMNPV genome were hybridized to Southern blots of restricted AcMNPV genome (LEISY et al. 1984). From the hybridization patterns, the authors were able to align the *Hind*III restriction map of OpMNPV with the *Hind*III and the *Pst*I restriction maps of AcMNPV and found that the cross-hybridizing regions of the genomes were arranged predominantly in a colinear manner.

Detailed homology analysis was also done on the genomes of CfMNPV and AcMNPV (ARIF et al. 1985). The AcMNPV used in these studies was described by LÜBBERT et al. (1981). Under low stringency conditions of hybridization (20% formamide, Tm-44° C), CfMNPV DNA hybridized to all the *Eco*RI fragments of AcMNPV genome except to *Eco*RI-R. When the cloned fragments of both genomes were used in the hybridization experiments, it was seen that the cloned *Eco*RI fragments of CfMNPV hybridized to the AcMNPV *Eco*RI map in approximately the same order as they appear on the CfMNPV restriction map (Fig. 2). This indicated that the gene order may be the same in both viruses. From these and the data already seen with the *Orgyia pseudotsugata* virus, it appears that the genomes of OpMNPV, CfMNPV, and AcMNPV are colinear. The close relationship among these three viruses was first mentioned by SMITH and SUMMERS (1982).

The data in Fig. 2 show that the fragment *Eco*RI-D of CfMNPV hybridized to widely separated segments on the AcMNPV genome, indicating some sequence translocation in the latter. Nonhybridizing regions such as the fragment

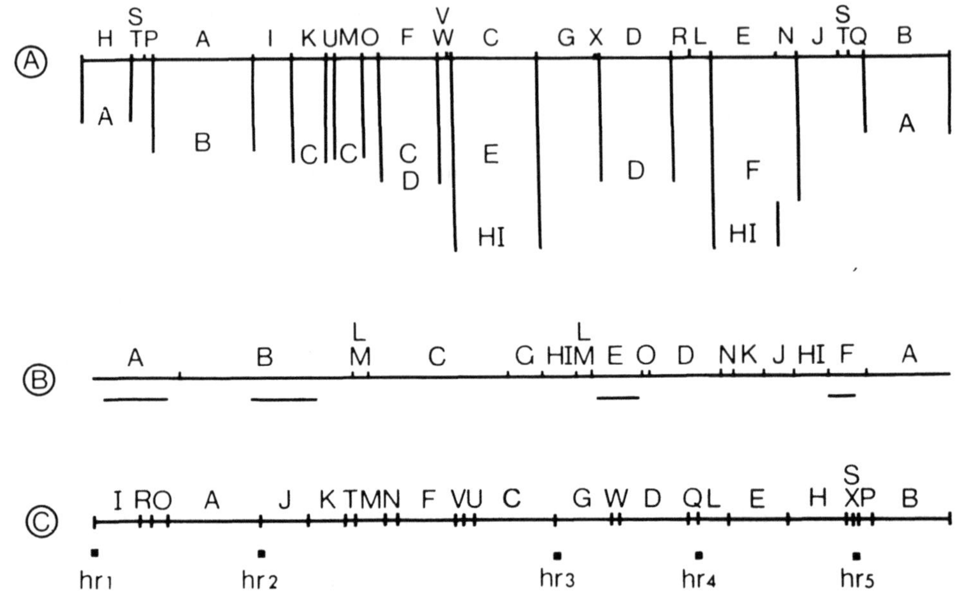

Fig. 2A–C. DNA sequence homologies between the CfMNPV and AcMNPV genomes. **A** The *Eco*RI restriction map of the AcMNPV described by LÜBBERT et al. (1981). The letters below this map refer to the CfMNPV *Eco*RI fragments that hybridize to that particular section of the AcMNPV genome. **B** The *Eco*RI restriction map of CfMNPV. The *bars below this map* indicate the locations of the reiterated sequences. **C** The *Eco*RI physical map of the AcMNPV genome described by COCHRAN and FAULKNER (1983), indicating the positions of the homologous repeats (hr)

*Eco*RI-R of AcMNPV were observed, similar to those reported by LEISY et al. (1984), who mentioned that such a region may be hypervariable or of independent origin. Figure 2 also shows that the four repeated sequences found in the CfMNPV genome correspond to four of the five homologous repeats of AcMNPV reported by COCHRAN and FAULKNER (1983).

5. Acquisition of Host DNA by the Viral Genome

The acquisition of cellular DNA sequences by viruses during the infectious cycle is a well-documented phenomenon (e.g., DEURING et al. 1981). This was also shown to occur in baculoviruses. MILLER and MILLER (1982) passaged AcNPV serially for 25 times in a *T. ni* cell line and isolated several mutants which produce few polyhedra plaques (FP). One such mutant (FP-D) contained new *Hind*III bands (a, b, and c) and lost the fragment *Hind*III-K of the parental genome. Band c was submolar but survived several plaque purifications. FP-D was also found to be a mixture of two viruses FP-DL, which was identical to FP-D, and FP-DS, which had lost the fragments *Hind*III-a, -b, and -K but with *Hind*III-c becoming molar. Physical maps of the DNAs showed that the fragments *Hind*III-a and -b were contiguous and constituted the fragment

*Hind*III-K. *Hind*III-c was actually the fragment *Hind*III-K with a 0.27-kb insertion. FP-DL contained a 7.3-kb insertion plus direct terminal repeats of 0.27 kb which the authors called TED. Differences in copy number and distribution of TED homologous sequences present in DNA isolated from the *T.ni* cell line as compared with the DNA isolated from larvae suggested that TED is a movable element.

Cellular DNA was also reported to be incorporated in the genomes of two closely related viruses, AcMNPV and *Galleria mellonella* NPV (GmMNPV). MP phenotype viruses were passaged successively three times in TN-368 cell line and FP mutants were isolated (FRASER et al. 1983). REN analyses showed that the most abundant mutation was the insertion of host sequences in the *Hind*III-J region of GmMNPV and in the *Hind*III-I of AcMNPV genomes. All host DNA insertions were located within the 500-bp segment between 35.0 and 37.7 map units in the AcMNPV genome, which indicated that, in these cases, recombination with cell DNA did not take place randomly over the entire viral genome (FRASER et al. 1983). The size of the inserted cell DNA varied from 0.8 to 2.8 kb, considerably smaller than that reported by MILLER and MILLER (1982). From semiquantitative data it was shown that one mutant from GmMNPV (GmFP1) hybridized more intensely with ^{32}P-labeled cell DNA probe than did the other FP mutants examined. This suggested hybridization to repetitive sequences in the cellular DNA. A second mutant (GmFP3) hybridized very weakly to the cell DNA probe and may, therefore, have originated from unique sequences. The mutant GmFP2, which exhibited intermediate homology to the labeled cellular DNA probe, contained sequences that were dispersed and moderately repetitive in the *T.ni* DNA. Most of the cellular DNA in GmFP2 genome represented a repeat sequence that contained a 1.28-kb *Alu*I fragment and may be considered an Alu family of repeats in the cell DNA. This may, or may not, bear similarity to the Alu families in mammalian genomes.

An interesting finding was that FP mutant-infected cells did not synthesize a 25K protein, normally present in MP-infected cells. There is a 25K protein which is coded for by a region within the fragment *Hind*III-J of AcMNPV but it remains to be seen if it is the same as the missing protein in FP-infected cells (VLAK et al. 1981; SMITH et al. 1982). FRASER et al. (1983) concluded that these cell DNA insertions appeared to be responsible for the FP phenotype, and the absence of the 25K protein may be a distinguishing feature of FP mutants. It should, however, be mentioned that some FP phenotypes may arise without detectable host DNA insertions in *Hind*-I of AcMNPV. Four FP mutants of AcMNPV, obtained after 25 serial passages, were analyzed for the presence of host DNA in the *Hind*III-I region. No host DNA was detected, indicating that other types of mutations were also responsible for the FP virus (POTTER and MILLER 1980).

Recently, five regions which may be associated with insertion of cellular DNA into the genome of SeMNPV were identified (BROWN et al. 1985). This is discussed further by D.L. KNUDSON in this volume.

Acknowledgments. I am grateful to Dr. John Spencer for supplying unpublished data on the sequence of *Hind*III-R of CfMNPV. I would also like to thank Miss Karen Griffiths for valuable editorial comments and Mr. Keith Brown for technical help.

References

Arif BM, Doerfler W (1984) Identification and localization of reiterated sequences in the *Choristoneura fumiferana* MNPV genome. EMBO J 3:525–529

Arif BM, Kuzio J, Faulkner P, Doerfler W (1984) The genome of *Choristoneura fumiferana* nuclear polyhedrosis virus: molecular cloning and mapping of the *Eco*RI, *Bam*HI, *Sma*I, *Xba*I and *Bgl*II restriction sites. Virus Res 1:605–614

Arif BM, Tjia ST, Doerfler W (1985) DNA homologies between the genomes of *Choristoneura fumiferana* and *Autographa californica* nuclear polyhedrosis viruses. Virus Res 2:85–94

Brown SE, Maruniak JE, Knudson DL (1984). Physical maps of SeMNPV baculovirus DNA: an AcMNPV genomic variant. Virology 136:235–240

Brown SE, Maruniak JE, Knudson DL (1985) Baculovirus (MNPV) genomic variants: characterization of *Spodoptera exempta* MNPV DNAs and comparison with other *Autographa californica* MNPV DNAs. J Gen Virol 66:2431–2441

Burand JP, Summers MD (1982) Alterations of *Autographa californica* nuclear polyhedrosis virus DNA upon serial passage in cell culture. Virology 119:223–229

Cochran MA, Faulkner P (1983) Location of homologous DNA sequences interdispersed at five regions in the baculovirus AcMNPV genome. J Virol 45:961–970

Crook NE, Spencer RA, Payne CC, Leisy DJ (1985) Variation in *Cydia pomonella* granulosis virus isolates and physical maps of the DNA from three variants. J gen Virol 66:2423–2430

Deuring R, Klotz G, Doerfler W (1981) An unusual symmetric recombinant between Adenovirus type 12 DNA and human cell DNA. Proc Natl Acad Sci 78:3142–3146

Fraser MJ, Smith GE, Summers MD (1983) Acquisition of host cell DNA sequences by baculoviruses: relationships between host DNA insertions and FP mutants of *Autographa californica* and *Galleria mellonella* nuclear polyhedrosis viruses. J Virol 47:287–300.

Harrap KA, Payne CC (1979) The structural properties and identification of insect viruses. Adv Virus Res 25:273–355

Huang YS, Hedberg M, Kawanishi CY (1982) Characterization of the DNA of a nonoccluded baculovirus, Hz-1V. J Virol 43:175–181

Kelly DC (1977) The DNA contained by nuclear polyhedrosis viruses isolated from four *Spodoptera* sp. (Lepidoptera, Noctuidae): genome size and homology assessed by DNA reassociation kinetics. Virology 76:468–471.

Kislev N (1985) Homology relationship between *Spodoptera littoralis* nuclear polyhedrosis virus and other baculoviruses. Intervirol 24:50–57

Knebel D, Lübbert H, Doerfler W (1985) The promotor of the late P-10 gene in the insect nuclear polyhedrosis virus *Autographa californica* activated by viral gene products and sensitivity to DNA methylation. EMBO J 4:1301–1306

Kuzio J, Faulkner P (1984) Regions of repeated DNA in the genome of *Choristoneura fumiferana* nuclear polyhedrosis virus. Virology 139:185–188

Kuzio J, Rohel DZ, Curry CJ, Krebs A, Carstens EB, Faulkner P (1984) Nucleotide sequence of the p10 polypeptide gene of *Autographa californica* nuclear polyhedrosis virus. Virology 139:414–418

Lee HH, Miller LK (1978) Isolation of genotypic variants of *Autographa californica* nuclear polyhedrosis virus. J Virol 27:754–767

Leisy DJ, Rohrmann GF, Beaudreau GS (1984) Conservation of genome organization in two multicapsid nuclear polyhedrosis viruses. J Virol 52:699–702

Lübbert H, Kruczek I, Tjia S, Doerfler W (1981) The cloned *Eco*RI fragments of *Autographa californica* nuclear polyhedrosis virus DNA. Gene 16:343–345

Maruniak JE, Brown SE, Knudson DL (1984). Physical maps of SfMNPV baculovirus DNA and its genomic variants. Virology 136:221–234

Miller DW, Miller LK (1982) A virus mutant with an insertion of a *copia*- like transposable element. Nature 299:562–564

Miller LK, Dawes KP (1979) Physical map of the DNA genome of *Autographa californica* nuclear polyhedrosis virus. J. Virol 29:1044–1055

Potter KN, Miller LK (1980). Correlating genetic mutations of a baculovirus with the physical map of the DNA genome. In: Fields BN, Jaenish R, Fox CF (eds) Animal virus genetics, vol 18 Academic, New York, pp 71–80

Rohrmann GF, Martingnoni ME, Beaudreau GS (1982) DNA sequence homology between *Autographa californica* and *Orgyia pseudotsugata* nuclear polyhedrosis viruses. J Gen Virol 62:137–145
Smith GE, Summers MD (1978) Analysis of baculovirus genomes with restriction endonucleases. Virology 89:517–527
Smith GE, Summers MD (1979) Restriction maps of five *Autographa californica* MNPV variants, *Trichoplusia ni* MNPV, and *Galleria mellonella* MNPV DNAs with endonucleases *Sma*I, *Kpn*I, *Bam*HI, *Sac*I, *Xha*I, and *Eco*RI. J Virol 30:828–838
Smith GE, Summers MD (1980) Restriction map of *Rachiplusia ou* and *Rachiplusia ou-Augotrapha californica* baculovirus recombinants. J Virol 33:311–319
Smith GE, Summers MD (1982) DNA homology among subgroup A, B, and C baculoviruses. Virology 123:393–406
Smith GE, Vlak JM, Summers MD (1982) *In vitro* translation of *Autographa californica* nuclear polyhedrosis virus early and late messenger RNA species. J Virol 44:199–208
Summers MD, Smith GE, Knell JD, Burand JP (1980) Physical maps of *Autographa californica* and *Rachiplusia ou* nuclear polyhedrosis virus recombinants. J Virol 34:693–703
Vlak JM, Smith GE, Summers MD (1981) Hybridization selection and *in vitro* translation of *Autographa californica* nuclear polyhedrosis virus in RNA. J Virol 40:762–771

The Regulation of Baculovirus Gene Expression

P.D. FRIESEN and L.K. MILLER[1]

1 Introduction

The most striking aspect of the infection process of occluded baculoviruses is the biphasic synthesis of two different infectious forms of the virus. In a typical cell, the virus first produces progeny extracellular viruses (EVs) to ensure its continued spread to other cells within the infected organism. Following EV synthesis, occluded virus (OVs) are produced that enable the virus to survive in the environment following the death of the infected organism and subsequently to initiate the primary infection of a new host [for previous reviews of the nature of baculoviruses see GRANADOS (1980), FAULKNER (1981), MILLER (1981a), KELLY (1982), CARTER (1984), MILLER (1984, 1986)].

The synthesis of the two forms of these viruses is temporally regulated and their sites of maturation are spatially distinct. Replication of the large (100- to 200-kb) circular, double-stranded DNA genome occurs in the nucleus, where it is also packaged into rod-shaped capsids. Initially, the nucleocapsids leave the nucleus and bud through the plasma membrane, thereby forming mature EV. Later, however, the nucleocapsids are sequestered in intranuclear membranes which are proliferated at late times in infection. The virus particles that are enveloped in the nucleus are then embedded within a crystalline protein matrix composed primarily (if not exclusively) of a single polypeptide known as either polyhedrin or granulin. Depending on the type of baculovirus, the

Dept. of Bacteriology and Biochemistry, University of Idaho, Moscow, ID 83843, USA

[1] *Current address:* Departments of Genetics and Entomology, University of Georgia, Athens, GA 30602, USA

mature occlusion bodies appear under the light microscope as either large (1–5 μm) polyhedra or small granules in the cases of nuclear polyhedrosis viruses and granulosis viruses respectively.

The transition from EV to OV synthesis is only one of many transitions that baculoviruses make in the course of their development. These developmental transitions appear to be modulated primarily by the regulation of viral gene expression. This review discusses the evidence for the regulation of gene expression by *Autographa californica* nuclear polyhedrosis virus (AcNPV) and the closely related variant *Trichoplusia ni* nuclear polyhedrosis virus (TnNPV) upon infection of their lepidopteran insect host cells in tissue culture.

Temporal control of gene expression upon NPV infection is demonstrated by SDS-polyacrylamide gel electrophoresis analysis of polypeptides synthesized at various times postinfection (pi) in cell cultures synchronously infected with EV. Early work identified at least 35 infected-cell-specific proteins (ICSP) synthesized during three phases of the AcNPV infection cycle (Carstens et al. 1979; Dobos and Cochran 1980; Wood 1980; Maruniak and Summers 1981). More recent work using metabolic inhibitors, temperature-sensitive (ts) mutants, and transient expression assays suggest that there may be four distinguishable phases of gene expression (Kelly and Lescott 1981; Miller et al. 1983; Wang and Kelly 1983; Gordon and Carstens 1984; Guarino and Summers 1986). The immediate early α-phase from approximately 2 to 4 h pi relies on host factors for expression while the delayed early or intermediate β-phase, from approximately 5 to 7 h pi, appears to be dependent on the synthesis of one or more α-phase proteins (Kelly and Lescott 1981; Guarino and Summers 1986). The early phase(s) precede viral DNA synthesis (Kelly and Lescott 1981; Miller et al. 1983; Wang and Kelly 1983; Gordon and Carstens 1984). The late γ-phase, from approximately 8 to 18 h pi, includes the bulk of viral DNA replication and infectious EV production. During this period, host protein synthesis is significantly curtailed (Kelly and Lescott 1981; Miller et al. 1983; Gordon and Carstens 1984). Commencing 18 h pi, the very late δ-phase is characterized by the prominent synthesis of the 29K polyhedrin protein, intranuclear envelopment of nucleocapsids, and occlusion body formation. By the time of cell lysis at approximately 70 h pi, polyhedrin has been so abundantly expressed that it constitutes approximately 20% of the total protein of the infected cell. Thus AcNPV displays a remarkably complex pattern of gene regulation. Recent work indicates that control at the transcriptional level plays a critical role in regulating gene expression in AcNPV-infected cells. This review discusses the nature of AcNPV transcriptional units and the cascade model for baculovirus gene regulation.

2 Organization and Temporal Regulation of AcNPV Transcription Units

Studies designed to elucidate the molecular mechanisms involved in the complex regulatory pathways of baculovirus replication are just beginning. Preliminary work has suggested that a large part of such regulation occurs at the level of transcription. Thus, a major thrust for many laboratories has been the determination of the transcriptional organization of AcNPV, the prototype of the nuclear polyhedrosis viruses. While progress has been rapid, deciphering the detailed organization of the large ~128-kb genome containing from 70 to 100 viral genes represents a formidable task. A number of different "state-of-the-art" techniques in molecular biology have been employed for this purpose. These include hybrid selection and in vitro translation of viral mRNA, Northern hybridization analysis, S1 nuclease and primer extension mapping of individual RNA transcripts, cDNA cloning, nucleotide sequencing, and others. Descriptions of these methodologies are readily available and will not be discussed in detail.

Thus far, several features of AcNPV transcriptional organization have been established. First, there is no clustering of RNA transcripts according to their temporal expression; early and late genes are distributed throughout the viral genome. Individual transcriptional units have been identified and are composed of overlapping RNAs expressed maximally during each of the temporal phases of infection (α, β, γ, and δ). The most abundant RNAs of several units are transcribed in the same direction and thereby form nested sets of RNAs with either a common 3′ or 5′ terminus. A frequent pattern of expression within units composed of RNAs with common 3′ ends involves the early transcription of the smallest RNAs followed later by subsequent synthesis of larger RNAs whose 5′ ends map further upstream. This unusual pattern appears to be the result of sequential activation of upstream promoters. RNA splicing has little, if any, role in the expression of the various genes of the overlapping nests. The experimental evidence for these conclusions is discussed below. Adopted AcNPV nomenclature (Vlak and Smith 1982) is used to designate the various restriction fragments of the viral genome mentioned.

2.1 Identification of Transcriptionally Active Regions of the AcNPV Genome

Early studies focused on mapping the location of regions of the viral genome that are transcriptionally active during early and late stages of infection. The first approach involved the hybridization of radiolabeled RNA isolated from infected cells to Southern blots of viral genomic DNA cleaved with various restriction enzymes. Hybridization of late (21 and 24 h pi) RNA indicated that transcription occurred from dispersed regions of the genome (Vlak et al. 1981; Vlak and Van der Krol 1982). While most, if not all, viral restriction fragments hybridized to late RNA, certain fragments showed more hybridization than others. This suggested that these regions of the genome were preferentially transcribed. Two of these regions, the *Hind*III-V/*Eco*RI-I and the *Hind*III-PQ/*Eco*RI-P fragments, were subsequently found to contain the δ-phase genes, polyhedrin (3.1–4.1 MU) and *p10* (88.5–89.3 MU) respectively; both are abundantly expressed late (24 h) after infection.

Using the same approach, other studies demonstrated that transcription occurring during the early stages of infection and before the onset of viral DNA replication (6 h pi) was localized to regions throughout the viral genome

(Erlandson and Carstens 1983). As infection proceeded, most of the viral genome was transcribed. It was noted that the regions encoding the late genes, polyhedrin and *p10*, were not transcribed at early times, an observation confirmed by Northern blot analysis of the RNAs transcribed from each of these genes (see below).

Further evidence that both early and late AcNPV genes are dispersed throughout the viral genome was provided by mapping various AcNPV proteins to the viral genome by hybrid selection and subsequent in vitro translation of viral mRNA (Vlak et al. 1981; Smith et al. 1982; Adang and Miller 1982; Esche et al. 1982). From four to six early viral polypeptides were identified in this manner. These were mapped to two different regions, 15–22 MU and 45–60 MU (Smith et al. 1982; Esche et al. 1982). Because of the limited sensitivity of the hybrid selection and translation procedures, these polypeptides probably represent only the most abundant of those expressed early. Certainly, Northern blot analysis of RNAs transcribed from other regions of the viral genome (see below) indicate that many more polypeptides, albeit less abundant, remain to be identified. In vitro translation of late viral mRNA resulted in the mapping of 10–24 additional viral proteins distributed throughout the genome. Interestingly, these studies also indicated that most, if not all, of the messenger-active RNA present in cells during the late stages of infection was viral specific (Smith et al. 1982; Esche et al. 1982; Rohel et al. 1983).

The construction of cDNA libraries to AcNPV poly(A)-containing [poly(A^+)] RNA has also provided information on the locations of actively transcribed regions of the viral genome. Of libraries prepared from RNA isolated early (2 h) and late (10 h) (Mainprize et al. 1986) and very late (27 h) (Adang and Miller 1982), cDNA clones have been identified and mapped to regions covering almost the entire genome. Of 18 early, 49 late, and 45 very late cDNAs examined, at least 20 different clones were identified. Assuming that the oligo (dT)-primed synthesis and subsequent cloning of viral cDNA was a random process, the frequency of cDNA within a specific library also reflects the relative abundance of the RNA transcript from which the clone was derived. As expected, the most abundant cDNAs identified in the very late (27 h) library were homologous to the polyhedrin and *p10* genes. Two clones were predominant within the 2 h early library and were homologous to the *Hind*III-F/*Eco*RI-I and *Hind*III-I/*Eco*RI-F regions respectively; only one cDNA predominated within the late 10 h library and was located within the *Hind*III-H/*Eco*RI-D region. Interestingly, the 2-h cDNA clones were also identified within the 10-h library and no occlusion-specific (27-h) cDNA clones were found early (Mainprize et al. 1986). The RNA transcripts homologous to several AcNPV cDNAs have been identified by Northern hybridization analysis (Friesen and Miller 1985; Mainprize et al. 1986) and are described below.

2.2 Temporal Regulation of AcNPV Transcription: General Aspects

Several laboratories have been involved in studies designed to examine the patterns of transcription occurring during the different stages of AcNPV replication. Northern blot hybridization, the primary technique used, has provided a catalogue of information on the number, size, relative abundance, and, in some cases, the temporal appearance and disappearance of RNAs mapping to different regions of the viral genome (SMITH et al. 1983; ROHEL et al. 1983; LÜBBERT and DOERFLER 1984; ROHEL and FAULKNER 1984; FRIESEN and MILLER 1985; ERLANDSON et al. 1985; RANKIN et al. 1986). Such data are prerequisites for more detailed analyses of the organization of individual transcription units and the molecular mechanisms involved in their regulation.

To date, well over 100 different RNA transcripts have been localized to the AcNPV genome. Not unexpected is that many more poly(A^+) transcripts have been identified than viral polypeptides detected via hybrid selection and in vitro translation (see above). Examination of the synthesis and accumulation of specific RNAs during the course of infection revealed that regulation of transcription is complex. RNAs present maximally during each of the four stages of replication (α, β, γ, and δ respectively) were identified (ROHEL and FAULKNER 1984; FRIESEN and MILLER 1985; ERLANDSON et. al. 1985; MAINPRIZE et al. 1986). Thus, analogous to viral protein synthesis, each phase in the infection cycle is associated with the appearance of various new RNAs. Temporal regulation of the RNAs within individual AcNPV transcription units are described in detail below. In addition to temporal regulation, transcription also appears to be regulated with respect to the relative abundance of the different RNAs. The two most abundant AcNPV RNAs are maximally transcribed during the very late δ-phase (24 h and on) and represent the mRNAs for polyhedrin and *p10* respectively (SMITH et al. 1983; ROHEL et al. 1983; ROHEL and FAULKNER 1984; FRIESEN and MILLER 1985). Another RNA (0.6-kb) transcribed very late and in amounts only slightly less than polyhedrin or *p10* RNAs was mapped to the *Hind*III-H/*Eco*RI-D region (M. WILSON, T. MAINPRIZE, P. FRIESEN, L. MILLER, submitted). The observed diversity in the relative abundance of these mRNAs and others transcribed earlier may reflect variation in the relative strength of promoter sequences located upstream of individual viral genes, copy number of the DNA template (genome), or transcript stability.

That various temporal classes of viral RNA have been identified during the course of infection supports the contention that regulation at the level of transcription is responsible, in part, for the synthesis of the various temporal phases of AcNPV proteins (see "Introduction"). This is also consistent with observation that the changes in the pattern of viral protein synthesis are closely correlated with the appearance of new viral RNAs as assayed by in vitro translation of hybrid-selected RNA (SMITH et al. 1982; ESCHE et al. 1982). Transcriptional regulation is especially apparent in the expression of the late δ-phase genes, polyhedrin and *p10*; the rapid synthesis of these proteins closely parallels the selectively high accumulation (transcription) of their respective mRNAs over a short period late in infection. Recent studies on the effects of protein and/or DNA synthesis inhibitors on viral transcription as well as the transient expression of individual viral genes indicate that the expression of certain viral genes require the proper expression of earlier viral genes. Evidence for regulatory cascades involved in the expression of AcNPV genes is discussed below.

2.3 Organization of AcNPV Transcription Units Containing Overlapping RNAs

Of key importance in gaining insight into the regulatory mechanisms of baculovirus gene expression is the construction of high-resolution transcription maps of individual viral genes. This includes the identification of regulatory regions involved in the apparent temporal specificity of transcription. To date, only a few viral RNAs have been mapped with respect to the direction of transcription and the location of either their 5′ or 3′ termini (SMITH et al. 1983; LÜBBERT and DOERFLER 1984; KUZIO et al. 1984; FRIESEN and MILLER 1985; RANKIN et al. 1986; GUARINO and SUMMERS 1986). Nevertheless, several organizational patterns have emerged. Most apparent is the existence of individual transcription units composed of an unusually large number of overlapping RNAs transcribed in the same direction and having either common 3′ or 5′ ends.

The prevalance of overlapping RNAs transcribed from the AcNPV genome was demonstrated by the use of cDNA clones as hybridization probes in Northern blot analyses of viral RNA isolated at various times after infection (MILLER et al. 1983; FRIESEN et al. 1985; MAINPRIZE et al. 1986). Of 17 cDNA clones derived from unique regions throughout the viral genome, all hybridized to more than one and, in some cases, up to 12 different-sized poly(A^+) RNAs. Overlapping RNAs synthesized maximally during each of the four temporal phases of replication were identified. The overlapping RNAs within several of these transcription units identified by the above cDNAs have been mapped (see below).

While the transcription of overlapping RNAs is a common pattern in the replication of AcNPV, derivation of these transcripts through RNA splicing is not. No evidence for splicing of either early or late viral RNAs has yet been obtained (SMITH et al. 1983; LÜBBERT and DOERFLER 1984a; FRIESEN and MILLER 1985). The possibility of RNA splicing late in infection was examined by LÜBBERT and DOERFLER (1984a). Late viral RNA (24 h pi) was hybridized to linearized AcNPV genomic DNA and treated with S1 nuclease as described by BERK and SHARP (1977). The resulting hybrids were then fractionated on two-dimensional (neutral followed by alkaline) agarose gels. The lack of large introns was demonstrated by the identical mobilities of the S1-resistant fragments in the neutral and alkaline directions. This method demonstrated extensive splicing of adenovirus RNA (LÜBBERT and DOERFLER 1984a). Splicing of minor species of RNA or splicing near the end of the RNAs could not be ruled out.

2.3.1 Regulation of Viral RNAs with a Common 3′ Terminus

Three different transcription units composed of overlapping RNAs with common 3′ ends have been mapped thus far to the AcNPV genome. These are designated here by the genomic restriction fragments to which they are homologous, the *Hind*III-I/*Eco*RI-F (I) and *Hind*III-B_2/*Eco*RI-H (B_2) units (LÜBBERT and DOERFLER 1984) and the *Hind*III-K/*Eco*RI-H, -S (K) unit (FRIESEN and

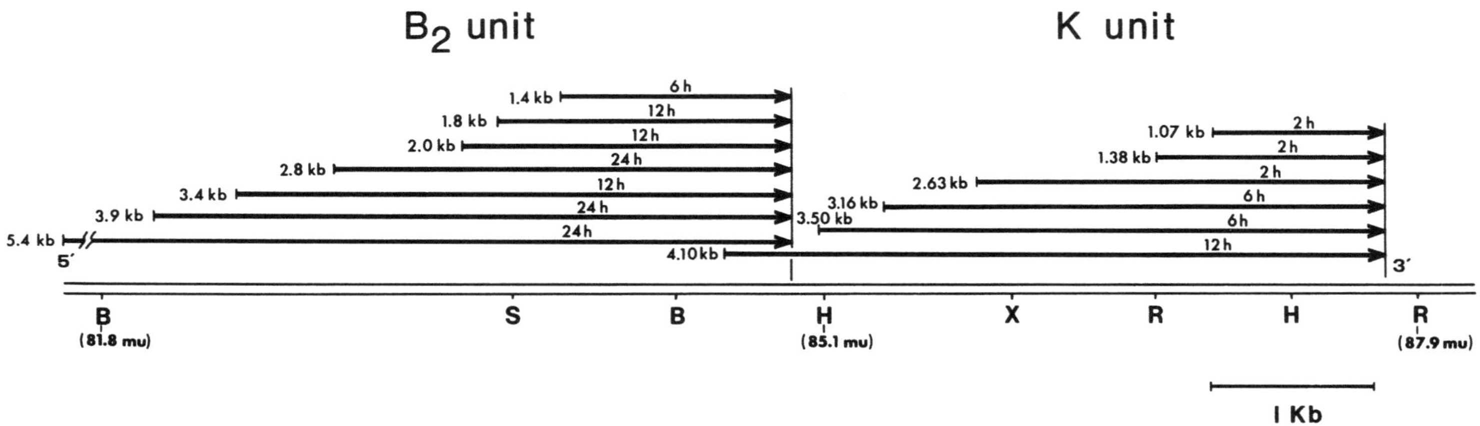

Fig. 1. Organization of overlapping 3′ coterminal transcripts mapping to the AcNPV *Hind*III-B_2 and *Hind*III-K regions. The AcNPV genome (from 81 to 88 MU) is transcribed from left to right (clockwise on the circular map) by two different groups (B_2 and K respectively) of overlapping RNAs having common 3′ termini. The size of each transcript including the poly(A) tract is designated in kilobases (kb). The time of appearance of each RNA in hours (h) after infection is also indicated (Friesen and Miller 1985; Mainprize et al. 1986). The 3′ end of the *Hind*III-B_2 transcription unit is located ~120 bp to the left of the *Hind*III restriction site at 85.1 MU (Lübbert and Doerfler 1984) while the 3′ end of the *Hind*III-K unit is located ~90 bp to the left of the *Eco*RI restriction site at 87.9 MU (Friesen and Miller 1985). At least one transcript (4.10 kb) originates within the *Hind*III-B_2 unit but terminates at the *Hind*III-K 3′ site as determined by S1-nuclease analysis (P.D. Friesen and L.K. Miller, unpublished data)

MILLER 1985). The latter two units are located adjacent to one another, extending from 81 to 88 MU on the AcNPV genome as illustrated in Fig. 1. Each transcription unit consists of a overlapping group of poly(A^+) RNAs, all transcribed from left to right (or clockwise on the circular AcNPV map) and terminated at a common 3′ site. The number and relative abundance of RNAs varies between groups. While only one transcript (4.10 kb) is indicated (Fig. 1), several RNAs originate within the *Hind*III-B_2 unit, extend through its stop signal, and finally terminate at the 3′ site used by RNAs of the *Hind*III-K unit (P.D. FRIESEN and L.K. MILLER, unpublished data).

Transcription of RNAs within these 3′ coterminal nests follows an unusual but common pattern. This pattern involves early transcription of the smallest RNAs followed later by transcription of successively longer RNAs mapping further upstream. The first RNAs to appear within the *Hind*III-K unit (Fig. 1) during infection are the three smallest (1.07, 1.38, and 2.63 kb respectively); maximum accumulation of the 1.07- and 1.38-kb RNAs occurs at 2 h while that of the larger 2.63-kb RNA occurs at 6 h (FRIESEN and MILLER 1985). The longest RNAs of the group are transcribed next; these appear first at 6 h, accumulate through 12 h, and decline thereafter (24 h). [*Note*: The *Hind*III-K transcripts were designated by FRIESEN and MILLER (1985) as α, β, and γ to indicate their periods of maximal expression 2, 6, and 12 h respectively rather than the various phases (α, β, and γ) of infection defined on the basis of protein synthesis; see below.] Thus as infection progresses, there is a subsequent increase in the size of RNAs transcribed within the unit.

High-resolution S1-nuclease mapping and primer extension analysis demonstrated that the overlapping *Hind*III-K RNAs were not spliced; this included the absence of small exons at the 5′ ends of the earliest 1.07- and 1.38-kb RNAs (FRIESEN and MILLER 1985). The data indicated that the two smallest RNAs are transcribed from promoters different from those used by the longer RNAs mapping upstream. Thus the observed pattern of successive replacement of early RNAs with longer RNAs transcribed later probably involves a sequential activation of different promoters located further and further upstream. Not yet ruled out is the possibility that each RNA within the unit is derived through a temporally regulated process involving post-transcriptional nucleolytic cleavage of a single, large precursor; however, such a process is unprecedented in eukaryotes.

The postulated early and late *Hind*III-K promoters differ in their responsiveness to yet undetermined factors encoded by the virus as demonstrated by the use of metabolic inhibitors (RICE and MILLER 1986). Infected cells treated with the protein synthesis inhibitor, cycloheximide, or the viral DNA synthesis inhibitor, aphidicolin, transcribed only the three early *Hind*III-K RNAs (1.07, 1.38, and 2.63 kb respectively). Transcription of the longer, late RNAs (3.16 kb and above) were blocked. This suggested that the three 1.07-, 1.38-, and 2.63-Kb RNAs (Fig. 1) represent immediate early transcripts (α-phase) since their synthesis does not require prior viral protein synthesis. The larger *Hind*III-K RNAs, on the other hand, represent late γ-phase transcripts since they are blocked by inhibitors of both viral protein and DNA synthesis. The *Hind*III-K unit therefore contains both immediate early α and late γ classes of RNA.

Transcriptional regulation within the two other 3′ coterminal units, *Hind*III-B_2 and *Hind*III-I/*Eco*RI-F respectively, is similar to that of the *Hind*III-K unit. Northern blot analysis using cDNA clones derived from both units indicated that the smallest of the overlapping RNAs within each group were transcribed early and were followed later by the sequential appearance of longer RNAs (MAINPRIZE et al. 1986). The pattern of sequential activation of upstream promoters may therefore be a common motif in baculovirus expression. Closer examination of the kinetics of transcription within the *Hind*III-B_2 unit revealed that the first RNA (1.4 kb) to appear did so at 6 h pi, 4 h later than the first RNAs of the adjacent *Hind*III-K unit (Fig. 1). Moreover, although the *Hind*III-B_2 RNAs appeared sequentially from the smallest to the largest (Fig. 1), maximum accumulation of all RNAs occurred very late (24 h or thereafter) in infection. To be determined is whether the *Hind*III-B_2 unit contains any immediate early α transcripts as does both the *Hind*III-K and *Hind*III-I units. The above data suggest that it does not. Thus similarly structured transcription units may contain different types of temporal promoters.

The functional significance of the overlapping organization of the 3′ coterminal transcription units is not clear. The overlapping arrangement may, for instance, provide variety in the expression of viral DNA accomplished by RNA splicing in other viral systems (LÜBBERT and DOERFLER 1984b). Key to understanding this fundamental problem in baculovirus expression will be the determination of the arrangement of viral genes encoded by each unit and the assignment of viral polypeptides to individual mRNAs. Since the 5′ ends of these poly(A^+) RNAs extend upstream from the termination site for various lengths up to 5 kb, each unit has the potential to code for several proteins some of which may share coding sequences. In vitro translation of mRNA hybrid-selected by either a cDNA clone (ADANG and MILLER 1982) or a genomic fragment (LÜBBERT and DOERFLER 1984b) of the *Hind*III-B_2/*Eco*RI-H region identified two to three polypeptides (34–37K, 26–28K, and 48K). Thus the *Hind*III-B_2 unit may encode more than one viral protein. The sequential appearance of different size classes of RNA during the various phases of viral replication within each of the above units also suggests that they direct the synthesis of several temporal classes of viral protein.

A possible *cis*-acting regulatory function for the overlapping arrangement of the 3′ coterminal transcription units is suggested by the dramatic turn on and turn off of RNAs within the *Hind*III-K unit. RNA polymerase molecules initiating at highly active promoters have been shown to block access of RNA polymerase to nearby downstream promoters through a process involving transcriptional interference or promoter occlusion (ADHYA and GOTTESMAN 1982; CULLEN et al. 1984). The concomitant increase in transcription from upstream *Hind*III-K promoters and decrease from the downstream promoters is consistent with promoter occlusion regulation. Thus, the longer late *Hind*III-K transcripts could repress transcription of the overlapping early genes as well as serve as messengers for unique late viral polypeptides. It should be noted, however, that repression of downstream transcription via a promoter occlusion mechanism is highly dependent upon the number and relative activity of upstream promoters and may therefore vary significantly between AcNPV transcription units.

2.3.2 Regulation of Late, Occlusion-Specific (δ) Genes

At least two very abundant RNA transcripts comprise the majority (greater than 90%) of the RNA present in cells late (24 h and on) after infection with AcNPV. These transcripts (1.28 kb and 0.72 kb) represent the mRNAs for proteins polyhedrin (29K) and *p10* (10K) respectively. Polyhedrin, first detected 12 h pi, is the major polypeptide component of the crystalline matrix of viral occlusion bodies assembled only late in the replication cycle. The function of the nonstructural protein *p10* is unknown. The complete nucleotide sequence of both AcNPV genes has been determined (HOOFT VAN IDDEKINGE et al. 1983; KUZIO et al. 1984). Because of the strict temporal specificity of transcription and the rapid accumulation of unusually high levels of RNA, both genes represent useful models for studies in eukaryotic transcriptional control.

Both the polyhedrin and *p10* genes are transcribed from left to right (clockwise) by several overlapping poly(A^+) RNAs (Fig. 2). Unlike the 3′ coterminal units described above, the *p10* and polyhedrin-related transcripts have common 5′ ends and different 3′ ends. The shortest transcript of each overlapping group (0.72 and 1.28 kb respectively) are the most abundant (SMITH et al. 1983; ROHEL et al. 1983; LÜBBERT and DOERFLER 1984a; ROHEL and FAULKNER 1984; FRIESEN and MILLER 1985; ERLANDSON et al. 1985; RANKIN et al. 1986). Relative abundance decreases with increasing length of the RNA.

A single transcriptional start site for the overlapping *p10* transcripts was identified by S1 nuclease analysis and mapped to a position 72–74 nucleotides (LÜBBERT and DOERFLER 1984) or 64 nucleotides (KUZIO et al. 1984) upstream from the *p10* translation initiation codon (AUG). A majority of the *p10* transcripts initiating from this position terminate approximately 540 nucleotides downstream, resulting in the major 0.72-kb RNA (RANKIN et al. 1986). At least two less abundant RNAs (2.56 and 2.72 kb) extend even further downstream and terminate near 90.5 MU within the *Hin*dIII-P/*Eco*RI-B region (Fig. 2) (LÜBBERT and DOERFLER 1984b; FRIESEN and MILLER 1985; RANKIN et al. 1986). These longer RNAs are apparently the result of transcriptional readthrough of several polyadenylation signals located at the end of the smaller 0.72-kb transcript. Consistent with this notion is the observation that the relative abundance of the *p10* and polyhedrin-related RNAs decrease as they extend downstream through additional termination signals (Fig. 2). A single transcription start site has also been identified 58 nucleotides upstream from the polyhedrin initiation codon (SMITH et al. 1983; HOOFT VAN IDDEKINGE et al. 1983). The overlapping 3.39- and 4.9-kb transcripts (Fig. 2) extend to the right of the major 1.28-kb polyhedrin transcript to as yet unmapped locations (FRIESEN and MILLER 1985, unpublished data; ERLANDSON et al. 1985 unpublished data).

Each of the 5′ coterminal RNAs within the *p10* and polyhedrin units respectively are transcribed with identical kinetics as demonstrated by Northern hybridization to cDNA clones (FRIESEN and MILLER 1985). The concomitant appearance of all RNAs within each unit supports the contention that transcription originates from a common promoter. The polyhedrin-related RNAs were detected as early as 6 h pi; however, maximum transcription did not occur until late in infection, 18–60 h. The observed early transcription of polyhedrin may

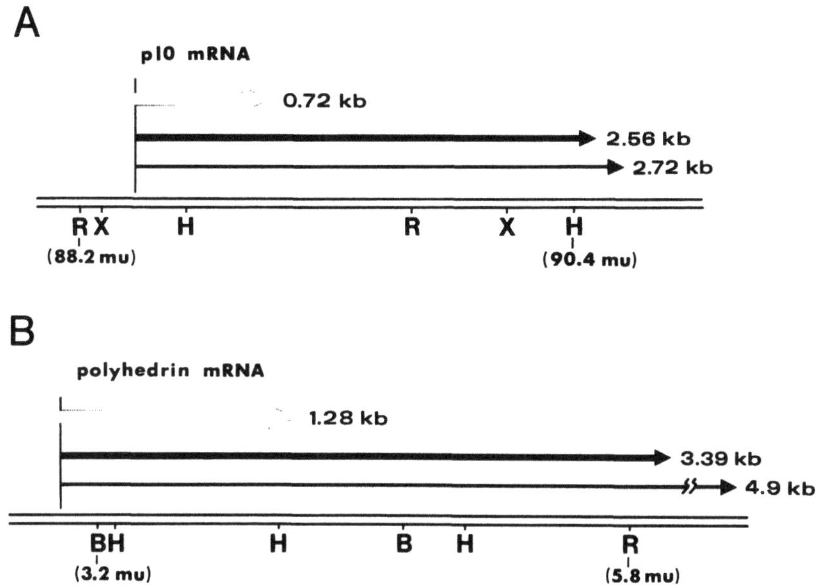

Fig. 2A, B. Transcriptional organization of two late δ-specific AcNPV genes, *p10* and polyhedrin. The *p10* **A** and polyhedrin **B** genes are transcribed from left to right by several overlapping RNAs with common 5′ ends. In both cases, the smallest RNAs, 0.72 kb and 1.28 kb respectively, represent the most abundant mRNAs for each gene. The longer, overlapping RNAs are less abundant as indicated by their respective line widths. These RNAs terminate downstream within the *Hind*III-P/*Eco*RI-B regions near 91 MU (LÜBBERT and DOERFLER 1984; FRIESEN and MILLER 1985; RANKIN et al. 1986). While the longer polyhedrin-related RNAs extend downstream from the major 1.28-kb RNA, the locations of their 3′ termini are tentative (FRIESEN and MILLER 1985, unpublished data; ERLANDSON et al. 1985, unpublished data)

represent low level constitutive expression which occurs prior to the major transcriptional activation modulated by viral-encoded factors. Transcription of both polyhedrin and *p10* was blocked by cycloheximide and aphidicolin (RICE and MILLER 1986) and indicated that prior viral expression was required for δ-specific expression.

The overlapping 5′ coterminal *p10* transcripts were not detected until 10–12 h pi but accumulated rapidly to high levels thereafter (ROHEL and FAULKNER 1984; FRIESEN and MILLER 1985; ERLANDSON et al. 1985). Transcription was detected as late as 60 h pi. Interestingly, two transcripts (1.1 and 1.5 kb) which overlap the *p10* gene from the upstream side were also identified (RANKIN et al. 1986). Both RNAs have a common 5′ end located upstream from the *p10* gene and terminate at positions located within (40 nucleotides) or at the end of the major 0.72-kb *p10* RNA. Both accumulate to maximum levels by 12 h pi but decline thereafter during the rapid increase in *p10* transcription. A single open reading frame was identified upstream from the *p10* gene (KUZIO et al. 1984; C. RANKIN and R. WEAVER, personal communication).

The functional significance, if any, of the 3′ extended ends of the 5′ coterminal polyhedrin and *p10* transcripts remains to be determined. No large (greater than 50 amino acids) open reading frames were identified immediately downstream from the *p10* gene within the same coding strand (KUZIO et al. 1984). Although these longer, nonspliced RNAs have the same protein-coding potential as the smallest RNA, their role in the regulation of overlapping downstream genes cannot be ruled out.

Recent analysis of the transcription of the polyhedrin and *p10* genes of the baculovirus *Orgyia pseudotsugata* MNPV revealed a striking similarity with the pattern of polyhedrin and *p10* transcription of AcNPV (LEISY et al. 1986a and 1986b). Thus it appears that regulation of the very late δ-genes maybe conserved between different species of baculoviruses.

3 A Regulatory Cascade in the Control of Baculovirus Gene Expression

One model for baculovirus gene regulation is a regulatory cascade in which the expression of each successive phase of polypeptide synthesis is dependent on one or more of the gene products of the previous phase. In such a model, the expression of β-phase genes is dependent on the expression of α-phase genes, the expression of one or more γ-phase genes is dependent on β-phase gene expression, and occlusion-specific δ-phase gene expression is dependent on γ-phase gene expression. Control of expression could be exerted at the transcriptional and/or translational level and could involve the expression of positive regulatory factors or the derepression of negative regulatory factors. It is likely that the large and complex baculoviruses employ a wide variety of regulatory mechanisms and that numerous regulatory cascades will be identified that make phase definition difficult. In this section of the review, the data supporting and defining a cascade model in the baculovirus infection process are surveyed along with a discussion of what types of regulatory mechanisms might be involved at each stage.

The distinction between early (α and β) and late (γ) genes and the dependence of γ gene expression on DNA replication is relatively well established. Early phase proteins and transcripts are characterized by the use of inhibitors of DNA synthesis or ts mutants defective in DNA synthesis. The first studies of this nature employed hydroxyurea, cytosine arabinoside (ara-C), and bromodeoxyuridine as inhibitors of baculovirus DNA synthesis (KELLY and LESCOTT 1981). In retrospect, however, none of these inhibitors are advisable for such studies. Hydroxyurea, an inhibitor of ribonucleotide reductase in at least some organisms, must be used in high concentration (800 μg/ml) to inhibit ICSP synthesis in baculovirus-infected insect cells. At elevated concentrations, this inhibitor blocks polypeptide synthesis at a very early phase, which KELLY and LESCOTT (1981) define as the α-phase; five ICSPs of 121K, 54K, 42K, 37K, and 29K are observed in hydroxyurea-blocked cells. Ara-C, however, blocked or delayed polypeptide synthesis at a preDNA synthesis stage (i.e., 6–7 h pi), which was referred to as the β-stage (KELLY and LESCOTT 1981). An explanation for why hydroxyurea blocks at an earlier stage than ara-C was not provided. The ara-C experiments are now obscured by the observation that ara-C is not an effective inhibitor of AcNPV DNA replication although it does effectively inhibit the uptake of exogenous [^{3}H]thymidine in infected cells (CARBONELL et al. 1985). At best, ara-C only delays viral DNA synthesis and the appearance of certain ICSPs (CARBONELL et al. 1985; see also DOBOS and COCHRAN 1980). Bromodeoxyuridine is a less potent suppressor (inhibitor) of polypeptide synthesis than hydroxyurea or ara-C (KELLY and LESCOTT 1981).

Further studies, however, have demonstrated more clearly that a block in DNA synthesis prevents late (γ) gene expression and the shut-off of host gene expression. Aphidicolin, an inhibitor of both the host α-polymerase and the viral-induced DNA polymerase, blocks viral gene expression between the early (α/β) and late (γ) phases (MILLER et al. 1981; WANG and KELLY 1983; RICE and MILLER 1986). Two temperature-sensitive (ts) mutants of AcNPV, tsB821 and ts8, are defective in viral DNA synthesis at the nonpermissive temperature (MILLER et al. 1983; GORDON and CARSTENS 1984). At the nonpermissive temperature, both mutants induce only early ICSPs and fail to switch off host protein synthesis. In the case of ts8, only three ICSPs of 30K, 31K, and 43K were observed (GORDON and CARSTENS 1984) whereas seven ICSPs of 145K, 125K, 115K, 72K, 69K, 62K, and 33K were observed during tsB821 infection (MILLER et al. 1983) at the nonpermissive temperature. Because of the variations in what proteins are considered ICSPs and the variability in molecular weight estimates from SDS-polyacrylamide gel electrophoresis analysis, an adequate comparison of the data from the three different laboratories is not possible. It is not clear from the existing literature what proteins belong to the α-phase, the β-phase, or both. Uniformly defining the proteins which characterize each phase will require more than merely SDS-polyacrylamide gel electrophoresis analysis in the future; specific antibodies will be useful reagents for this purpose.

Based on the fact that the seven ICSPs induced by tsB821 were synthesized by 2–4 h pi, MILLER et al. (1983) considered the block in tsB821 to reside in an α gene essential for β-phase synthesis. However, it is also possible that the tsB821 block is at the β-stage; either way, the block appears to be in an early function that can block late ICSP synthesis and DNA synthesis. GORDON and CARSTENS (1984) also consider that the ts8 mutation interferes with an α function. They base their conclusion on a series of temperature shift experiments that indicate that the effect of the ts8 mutation is not immediately exerted following the shift to or from the nonpermissive temperature and suggest that the ts8 mutation affects the synthesis of β-phase proteins required for viral DNA synthesis. The ts8 mutation may affect the synthesis of an α protein involved in β-phase expression or it may affect the conformation of a β-phase protein (during its translation) that is involved in replication. It is interesting to note that ts8 and tsB821 have been mapped to two different locations on the physical map of AcNPV; ts8 maps between 60.1 and 62 MU whereas tsB821 maps between 90.7 and 1.9 MU (MILLER 1981 b; GORDON and CARSTENS 1984). Regardless of the details of whether the ts defects affect α- or β-phase expression, the work with these mutants shows that early gene products are necessary for DNA synthesis and for the shift from the early phase to the late phase. The work with aphidicolin indicates that abundant late gene expression is dependent on DNA synthesis.

Support for two distinct phases of gene expression prior to DNA synthesis (6–7 h pi) was suggested from protein synthesis blocking experiments using the inhibitor cycloheximide from the beginning of infection to 8 or 14 h pi (KELLY and LESCOTT 1981). A specific set of five ICSPs (the α set, vide supra) were synthesized immediately following release from the block; synthesis of the β set of ICSPs was delayed and was followed by γ-phase synthesis. The later the cycloheximide block was applied (1–8 h pi, 2–8 h pi, or 4–8 h pi), the more rapid the onset of β-phase ICSP. Using amino acid analogs (azetidine, canavanine, and *p*-fluorophenylalanine) to block the synthesis of functional proteins at various times pi, further indications of a distinct α- and β-phase were obtained. A block from 1 to 8 h pi delayed the synthesis of β and γ peptides (KELLY and LESCOTT 1981). KELLY and LESCOTT (1981) further concluded that inhibition of any stage prevented the subsequent stage from occurring and that γ ICSP synthesis may control the switch off of the α- and β-phase. Considering the importance of these observations to the cascade model of gene regulation, confirmation of these results by other laboratories and clearer definition of α- and β-phase proteins would be appropriate.

Evidence for a cascade model involving both immediate early (α) and delayed early (β) phases also comes from studies involving the transient expression of plasmid DNAs carrying a chloramphenical acetyl transferase (CAT) reporter gene under the control of the promoter for a 39K "early" protein which is encoded by the 20.7- to 22.7-MU region of AcNPV (GUARINO and SUMMERS 1986). Expression of CAT from the 39K promoter requires cotransfection of the plasmid DNA carrying the 39K/CAT fusion with fragments of viral DNA containing the region from 95.0 to 96.9 MU on the AcNPV physical map. This region contains a gene, referred to as the immediate early gene number one (IE-1), which can be expressed by host factors alone. Immediate early expression of IE-1 was demonstrated by constructing a plasmid, pIE-1/CAT, containing the CAT gene under the control of the IE-1 promoter; transfection of cells with this plasmid alone resulted in CAT gene expression. Transfection with a plasmid containing both the IE-1 gene and the CAT gene under 39K promoter control resulted in CAT gene expression equivalent to the activity induced by cotransfection of two plasmids carrying the two genes separately (GUARINO and SUMMERS 1986). It may be noteworthy that IE-1 maps in the general location of the tsB821 mutation.

These transient expression experiments provide a clear demonstration of a regulatory cascade operative in baculovirus gene expression. The experiments also suggest that an immediate early (α) and delayed early (β) phase exist in at least one baculovirus regulatory pathway. However, it remains to be formally demonstrated that the 39K protein is an "early" protein and that its synthesis is not substantially blocked by inhibitors of DNA synthesis. A 1.1-kb transcript encoded by the *Hind*III-M/*Eco*RI-J region (20.4–22.2 MU) of AcNPV appears to be dramatically enhanced late in infection (MAINPRIZE et al. 1986); it remains to be determined if this transcript encodes the 39K protein. If the 39K protein is actually a late protein, then the existence of two distinct early phases (α and β) is questionable. Nevertheless, it is clear that IE-1 encodes a *trans*-acting factor which stimulates 39K gene expression and it will be interesting to determine if the IE-1 gene product acts at the transcription or translational level.

The regulatory cascade of AcNPV appears to involve, at least in part, control at the transcriptional level with the sequential activation and, in some cases, the deactivation of promoters. The *Hind*III-K region of AcNPV (Fig. 1) exhibits three temporal classes of transcripts which have been defined by the time of their maximal abundance (Friesen and Miller 1985, vide supra). The two smallest transcripts of *Hind*III-K (1.07 and 1.38 kb) are maximally present at 2 h pi, a larger 2.63-kb transcript is maximally present at 6 h pi, while the two longest transcripts, 3.16 and 3.50 kb, are maximally present at 12 h pi. These studies indicate that transcriptional control plays an important role in gene regulation during the early and late phases of infection. Numerous other examples of transcriptional regulation have been provided by Northern blot analysis of poly(A^+) RNAs isolated at various times pi using viral cDNAs as probes (Friesen and Miller 1985; Mainprize et al. 1986). The correspondence between maximal RNA levels observed by Northern blot analysis and α-, β-, and γ-phase protein synthesis remains to be demonstrated.

Recent research has indicated that both protein and DNA synthesis inhibitors (cycloheximide and aphidicolin, respectively) block only the synthesis of the latest class of *Hind*III-K transcripts (Rice and Miller 1986). The cycloheximide results indicate that the *Hind*III-K 1.07-, 1.38-, and 2.63-kb transcripts (Fig. 1) are all immediate early α transcripts since they require for their synthesis only host proteins. It is not yet known if the proteins that may be encoded by these immediate-early transcripts are α, β, or possibly late proteins since some translational control mechanism may be operative in regulating their final expression. It is therefore important to distinguish α-, β-, and γ-phase transcripts from α-, β-, and γ-phase proteins. It is likely, however, that proteins encoded by the smallest *Hind*III-K transcripts are early since the levels of these transcripts decline significantly at late times.

The difference observed in the time of maximal abundance of the two immediate early (1.07-kb and 2.63-kb) *Hind*III-K transcripts (2 h pi vs. 6 h pi respectively) could be due to α-phase protein repression of the 1.07-kb transcript or α-phase protein stimulation of 2.63-kb transcript. Support for the former hypothesis is the observation that both transcripts accumulate to substantial levels at 24 h pi in the presence of cycloheximide (Rice and Miller 1986). More research will be necessary to determine if there exists a class of AcNPV transcripts that are inhibited with cycloheximide but synthesized in the presence of aphidicolin; such transcripts would represent a transcriptionally controlled β-class of transcripts. It may be found that the abundance of some transcripts is continually regulated by gene products of each successive phase, thus making rigid definitions of α-, β-, γ- and δ-phases obsolete.

The latest of the *Hind*III-K transcripts (3.16, 3.50, and 4.10 kb) are γ transcripts since they are maximally present at 12 h pi (Friesen and Miller 1985) and are dependent on DNA synthesis for their detectable expression. In the presence of the DNA replication inhibitor, aphidicolin, the synthesis of the γ transcripts is blocked (Rice and Miller 1986). The protein products of the *Hind*III-K γ transcripts are probably expressed as γ-phase proteins since the transcripts are maximally abundant at 12 h pi and decline during the very late occlusion phase.

Occlusion-specific (δ-phase) genes are those maximally expressed after 18 h pi. The control of expression of the most prominent occlusion-specific proteins, polyhedrin and *p10*, is known to involve transcriptional control (vide supra) although translational control mechanisms may also be involved. Synthesis of these transcripts may be detected as early as 6–12 h pi using high specific activity probes, but they do not become abundant until 24 h pi (MILLER et al. 1983; FRIESEN and MILLER 1985). After 24 h, the polyhedrin and *p10* RNAs become the predominant transcripts of infected cells (ADANG and MILLER 1982; ROHEL et al. 1983; SMITH et al. 1983).

It remains to be demonstrated formally that the accumulation of these transcripts at very late times is due to selective high transcriptional initiation at the promoters of these genes or whether it is due to selective stability of the mRNAs in the cell. Given the very rapid accumulation of these transcripts over a short period (6 h parentheses, it is likely that stimulation of transcriptional initiation plays a role. The rapid decline of γ-phase transcripts (e.g., *Hind*III-K γ transcripts) between 12 and 24 h pi may be due to general destabilization of all except a select few mRNAs. The fact that the most highly conserved DNA sequences upstream of the coding sequences of *p10* and polyhedrin genes of AcNPV and other baculoviruses lie within the 5′ leader sequence of the mRNAs (VLAK and ROHRMANN 1985) suggests any number of possibilities for gene regulation: unique promoter placement, enhanced stability of the mRNA, and/or enhanced translational capability.

Transient expression assays of a plasmid carrying a *p10* promoter/CAT gene fusion have shown that significant expression of the *p10* promoter gene requires other viral gene products (KNEBEL et al. 1985). This is not an unexpected observation given the very late expression of this gene and the probable need to express genes of the α-, β- and γ-phases to allow δ-gene expression. The *p10* and polyhedrin genes are probably transcribed by the viral-induced, α-amanitin-resistant RNA polymerase II-like activity described by Weaver and his colleagues (GRULA et al. 1981; FUCHS et al. 1983).

By 18 h pi, the major portion of RNA synthesized in isolated nuclei is viral and a major portion is resistant to α-amanitin (FUCHS et al. 1983). It remains to be determined conclusively whether the viral-induced α-amanitin-resistant polymerase activity transcribes only occlusion-specific genes (particularly polyhedrin and *p10*) or whether it is also involved in the expression of γ-phase genes. Since a substantial increase in the proportion of α-amanitin-resistant RNA synthesis is observed between 8 and 18 h pi, it is probable that the viral-induced RNA polymerase activity is involved in both γ- and δ-phase gene expression (FUCHS et al. 1983). There are several possible explanations for the observed leveling off of α-amanitin-resistant, viral-specific RNAs observed by FUCHS et al. (1983) at 24 h pi including the possible saturation of *p10* and polyhedrin sequences immobilized on the filters used to detect viral sequences. Thus virtually all RNA synthesis at 24 h pi and thereafter may be mediated by this novel polymerase activity. Given that the bulk of viral RNA made at 8 h pi is sensitive to α-amanitin, it is unlikely that this polymerase plays a role in β-phase transcription. The new RNA polymerase probably is involved in the switch from early to late (β to γ) gene expression and an additional factor or factors are required for the transition for the γ- to δ-phase gene expression.

The possible role of the nucleoprotein structures of viral and host DNAs in transcriptional regulation has not been determined. A dramatic alteration in the nucleoprotein structure of viral DNA occurs approximately 24 h pi (WILSON and MILLER 1986). During the early (α and β) and late (γ) phases of infection, viral DNA adopts a nucleoprotein structure which is nucleosome-like as determined by micrococcal nuclease digestion. By 24 h pi, however, the entire viral genome, including transcriptionally active δ-phase viral genes, adopts a unique nucleoprotein structure. This novel nucleoprotein structure is characterized by the generation of DNA fragments of 90 and 120 bps in length upon micrococcal nuclease digestion. The adoption of this new "subnucleosomal" structure is accompanied by the association of a basic 15K protein in nuclear chromatin preparations. Since this 15K protein is also a virus structural protein, its role in δ-phase gene expression remains unclear. The "subnucleosomal" structure may result from the packaging of viral DNA during the viral maturation process.

Interestingly, host DNA remains in a nucleosomal structure throughout the course of infection (WILSON and MILLER 1986). This observation raises questions regarding the compartmentalization of viral DNA at a late time in infections. Another interesting observation is the association of a viral-induced 39K protein with chromatin at late times ($\sim$10 h pi). It is not known if this protein is associated with viral or host DNA. It could potentially play a role in the turn-off of host protein synthesis that occurs during the γ-phase viral gene expression or the expression of γ- and δ-phase viral genes.

The recent observation that transcription of a cellular gene (SF21) is induced upon AcNPV infection of *Spodoptera frugiperda* cells (T. MAINPRIZE, K-J. LEE, and L.K. MILLER, unpublished work) makes it important to maintain the distinction between infected-cell-specific proteins (ICSP) and viral protein synthesis. The possibility that host gene products are involved in the regulation of viral gene expression remains an important consideration. In the case of SF21, the transcripts are maximally present at 12 h pi, similar to the *Hind*III-K γ transcripts described by FRIESEN and MILLER (1985).

References

Adang MJ, Miller LK (1982) Molecular cloning of DNA complementary to mRNA of the baculovirus *Autographa californica* nuclear polyhedrosis virus: location and gene products of RNA transcripts found late in infection. J Virol 44:782–793

Adhya S, Gottesman M (1982) Promoter occlusion: transcription through a promoter may inhibit its activity. Cell 29:939–944

Berk AJ, Sharp PA (1977) Sizing and mapping of early adenovirus mRNAs by gel electrophoresis of S1 endonuclease-digested hybrids. Cell 12:721–732

Carbonell LF, Klowden MJ, Miller LK (1985) Baculovirus-mediated expression of bacterial genes in dipteran and mammalian cells. J Virol 56:153–160

Carstens FB, Tjia ST, Doerfler W (1979) Infection of *Spodoptera frugiperda* cells with *Autographa californica* nuclear polyhedrosis virus. I. Synthesis of intracellular proteins after virus infection. Virology 99:386–398
Carter JB (1984) Viruses as pest-control agents. In: Russell GE (ed) Biotechnology and genetic engineering reviews, vol. I Intercept, Newcastle-upon-Tyne, pp 375–419
Cullen BR, Lomedico PT, Ju G (1984) Transcriptional interference in avian retroviruses: implications for the promoter insertion model of leukaemogenesis. Nature (London) 307:241–245
Dobos P, Cochran MA (1980) Protein synthesis in cells infected by *Autographa californica* nuclear polyhedrosis virus (AcNPV): the effect of cytosine arabinoside. Virology 103:446–464
Erlandson MA, Carstens EB (1983) Mapping early transcription products of *Autographa californica* nuclear polyhedrosis virus. Virology 126:398–402
Erlandson MA, Gordon J, Carstens EB (1985) Size and map locations of early transcription products on the *Autographa californica* nuclear polyhedrosis virus genome. Virology 142:12–23
Esche H, Lübbert H, Siegmann B, Doerfler W (1982) The translation map of the *Autographa californica* nuclear polyhedrosis virus (AcNPV) genome. EMBO 1:1629–1633
Faulkner P (1981) Baculovirus. In: Davidson EW (ed) Pathogenesis of invertebrate microbial diseases. Allanheld, Osmun, Totowa, NJ, pp. 3–37
Friesen PD, Miller LK (1985) Temporal regulation of baculovirus RNA: overlapping early and late transcripts. J Virol 54:392–400
Fuchs LY, Woods MS, Weaver RF (1983) Viral transcription during *Autographa californica* nuclear polyhedrosis virus infection: a novel RNA polymerase induced in infected *Spodoptera frugiperda* cells. J Virol 48:641–646
Gordon JD, Carstens EB (1984) Phenotypic characterization and physical mapping of a temperature-sensitive mutant of *Autographa californica* nuclear polyhedrosis virus defective in DNA synthesis. Virology. 138:69–81
Granados RR (1980) Infectivity and mode of action of baculoviruses. Biotechnology and Bioengineering 22:1377–1405
Grula MA, Buller PL, Weaver RF (1984) α-Amanitin resistant viral RNA synthesis in nuclei isolated from nuclear polyhedrosis virus-infected *Heliothis zea* larvae and *Spodoptera frugiperda* cells. J Virol 38:916–921
Guarino LA, Summers MD (1986) Functional mapping of a *trans*-acting gene required for expression of a baculovirus delayed early gene. J Virol 57:563–571
Hooft van Iddekinge BJL, Smith GE, Summers MD (1983) Nucleotide sequence of the polyhedrin gene of *Autographa californica* nuclear polyhedrosis virus. Virology 131:561–565
Kelly DC (1982) Baculovirus replication. J Gen Virol 63:1–13
Kelly DC, Lescott T (1981) Baculovirus replication: protein synthesis in *Spodoptera frugiperda* cells infected with *Trichoplusia ni* nuclear polyhedrosis virus. Microbiologica 4:35–37
Knebel D, Lübbert H, Doerfler W (1985) The promoter of the late *p10* gene in the insect nuclear polyhedrosis virus *Autographa californica*: activation by viral gene products and sensitivity to DNA methylation. EMBO 4:1301–1306
Kuzio J, Rohel DZ, Curry CJ, Krebs A, Carstens EB, Faulkner P (1984) Nucleotide sequence of the *p10* polypeptide gene of *Autographa californica* nuclear polyhedrosis virus. Virology 139:414–418
Leisy DJ, Rohrmann GF, Nesson M, Beaudreau GS (1986a) Nucleotide sequencing and transcriptional mapping of the *Orgyia pseudotsugta* multicapsid nuclear polyhedrosis virus *p10* gene. Virology 153:157–167
Leisy DJ, Rohrmann GF, Beaudreau GS (1986b) The nucleotide sequence of the polyhedrin gene region from the multicapsid baculovirus of *Orgyia pseudotsugata*. Virology 153:280–288
Lübbert H, Doerfler W (1984a) Mapping of early and late transcripts encoded by the *Autographa californica* nuclear polyhedrosis virus genome: is viral RNA spliced? J Virol 50:497–506
Lübbert H, Doerfler W (1984b) Transcription of overlapping sets of RNAs from the genome of *Autographa californica* nuclear polyhedrosis virus: a novel method for mapping RNAs. J Virol 52:255–265
Mainprize TM, Lee K-J, Miller LK (to be published) Variation in the temporal expression of overlapping baculovirus transcripts. Virus Research
Maruniak JE, Summers MD (1981) *Autographa californica* nuclear polyhedrosis virus phosphoproteins and synthesis of intracellular proteins after virus infection. Virology 109:25–34

Miller LK (1981a) A virus vector for genetic engineering in invertebrates. In: Panopoulos NJ (ed) Genetic engineering in the plant sciences. Praeger, New York, pp 203–224
Miller LK (1981b) Construction of a genetic map of the baculovirus *Autographa californica* nuclear polyhedrosis virus by marker rescue of temperature-sensitive mutants. J Virol 39:973–976
Miller LK (1984) Exploring the gene organization of baculoviruses. In: Maramorosch K, Koprowski H (eds) Methods in virology, vol. VII. Academic, New York, pp 227–258
Miller LK (to be published) The genetics of baculoviruses: In: Granados RR, Federici BA (eds) The biology of baculoviruses. CRC, Boca Raton, Florida
Miller LK, Jewell JE, Browne D (1981) Baculovirus induction of a DNA polymerase. J Virol 40:305–308
Miller LK, Miller DW, Adang MJ (1983) An insect virus for genetic engineering: developing baculovirus polyhedrin substitution vectors. In: Lurquin PF, Kleinhofs A (eds) Genetic engineering in eukaryotes Plenum New York, pp 89–97
Miller LK, Trimarchi RE, Browne D, Pennock GD (1983) A temperature-sensitive mutant of the baculovirus *Autographa californica* nuclear polyhedrosis virus defective in an early function required for further gene expression. Virology 126:376–380
Rankin, Ladin CB, Weaver R (1986) Physical mapping of temporally regulated, overlapping transcripts in the region of the 10K protein gene in *Autographa californica* nuclear polyhedrosis virus. J Virol 57:18–27
Rice WC, Miller LK (to be published) Baculovirus transcription in the presence of inhibitors and nonpermissive *Drosophila* cells. Virus Research
Rohel DZ, Faulkner P (1984) Time course analysis and mapping the *Autographa californica* nuclear polyhedrosis virus transcripts. J Virol 50:739–747
Rohel DZ, Cochran MA, Faulkner P (1983) Characterization of two abundant mRNAs of *Autographa californica* nuclear polyhedrosis virus present late in infection. Virology 124:357–365
Smith GE, Vlak JM, Summers MD (1982) *In vitro* translation of *Autographa californica* nuclear polyhedrosis virus early and late mRNAs. J Virol 44:199–208
Smith GE, Vlak JM, Summers MD (1985) Physical analysis of *Autographa californica* nuclear polyhedrosis virus transcripts for polyhedrin and 10,000-molecular-weight protein. J Virol 45:215–225
Vlak JM, Krol S van der (1982) Transcription of the *Autographa californica* nuclear polyhedrosis virus genome: location of late cytoplasmic mRNA. Virology 123:222–228
Vlak JM, Rohrmann GF (1985) Nature of polyhedrin. In: Maramorosch K, Sherman K (eds) Viral insecticides and biological control. Academic, New York
Vlak JM, Smith GE (1982) Orientation of the genome of *Autographa californica* nuclear polyhedrosis virus: a proposal. J Virol 41:1118–1121
Vlak JM, Smith GE, Summers MD (1981) Hybridization selection and *in vitro* translation of *Autographa californica* nuclear polyhedrosis virus mRNA. J Virol 40:762–771
Wang X, Kelly DC (1983) Baculovirus replication: purification and identification of the *Trichoplusia ni* nuclear polyhedrosis virus-induced DNA polymerase. J Gen Virol 64:2229–2236
Wilson ME, Miller LK (1986) Changes in the nucleoprotein complexes of a baculovirus DNA during infection. Virology 151:315–328
Wood HA (1980) *Autographa californica* nuclear polyhedrosis virus-induced proteins in tissue culture. Virology 102:21–27

Expression of the *Autographa californica* Nuclear Polyhedrosis Virus Genome in Insect Cells: Homologous Viral and Heterologous Vertebrate Genes – The Baculovirus Vector System

W. DOERFLER

1 Introduction

Investigations on the insect baculovirus *Autographa californica* nuclear polyhedrosis virus (AcNPV) have produced a wide spectrum of problems related to the molecular biology of insect cells. Since this virus system has been studied in considerable detail, one can begin to use it as a probe into molecular mechanisms in insect cells. In the first part of this chapter, I shall concentrate on a discussion of the modes of expression of the AcNPV genome which are characterized by several peculiar features:

(1) As in other large viral DNA genomes, there is temporal regulation of viral DNA expression leading to the production of at least four classes of RNA, immediate early, early, late, and very late.
(2) There is no indication to date that viral transcripts can be spliced.

Institute of Genetics, University of Cologne, Weyertal 121, D-5000 Cologne 41

Current Topics in Microbiology and Immunology, Vol. 131

(3) In several parts of the viral genome, overlapping sets of RNAs with common 3′- or 5′-termini have been detected whose significance has not yet been clarified.
(4) There is evidence that a virus-induced or a modified host DNA-dependent RNA polymerase is responsible for the transcription of late viral functions.

The expression of the AcNPV genome in insect cells has lately assumed a dimension of considerable practical importance in that the AcNPV genome has been developed to one of the highly efficient eukaryotic expression vector systems, which allows the expression of many eukaryotic (and prokaryotic) genes in insect cells (SMITH et al. 1983a). I am going to summarize this application of the AcNPV genome in the second part of this review. The vector system is based on the use of one of the strong late promoters of the AcNPV genome, the promoter of the polyhedrin gene. Numerous foreign genes, – e.g., the genes for human β-interferon, α-interferon, *Escherichia coli* β-galactosidase, human *c-myc* protein, human interleukin 2, influenza virus hemagglutinin, influenza virus polymerase complex, gene products of lymphocytic choriomeningitis arena-virus, and rotavirus capsid antigens – have been fused to that promoter and recombined into the AcNPV genome. Insect cells infected with the recombinant virus not only synthesized the foreign proteins under the orchestration of the polyhedrin promoter, but modified these proteins post-translationally so that biologically active products were generated. It is a safe assumption that, in addition to the proteins on which published work is available, a host of bio-medically interesting polypeptides have been expressed with the aid of this vector system. Information on some of these endeavors has not been published.

Four major areas can be listed that document the attractiveness of the AcNPV system as an area of research in molecular biology.
(1) AcNPV has served as an excellent model to elucidate mechanisms in the molecular biology of insect cells.
(2) AcNPV has been rendered one of the most versatile eukaryotic expression vectors. The potential of this system has been expanded by the demonstration that *Bombyx mori* larvae can be infected with recombinant AcNPV and express large amonts of the foreign protein (α-interferon) (MAEDA et al. 1985).
(3) Traditionally, baculoviruses have been considered useful agents in insect pest control. This widely publicized aspect of baculovirus work will not be treated in this review.
(4) Recently, it has been suggested (D.H. BISHOP, personal communication; NEWMARK 1986) that a genetically manipulated AcNPV genome could be used to test the fate and/or the stability of a manipulated viral genome which would be released into nature, in this case in the forests of the Scottish highlands. In this "virus-banding" (in analogy to bird-banding) experiment, a biologically innocuous nucleotide sequence, which is studded with stop codons, will become part of the AcNPV genome. After dissemination of this recombinant baculovirus among natural populations of lepidopteran larvae, the stability of this "useless" DNA sequence in the natural AcNPV progeny can be assessed.

2 Expression of the AcNPV Genome in Insect Cells

2.1 The Translational Map of the AcNPV Genome

Several laboratories have contributed translational maps of the AcNPV genome by assigning map locations of AcNPV genome-encoded polypeptides on the restriction map (ADANG and MILLER 1982; SMITH et al. 1982; ESCHE et al. 1982; ROHEL and FAULKNER 1984). Early (6 h) or late (24 h) after the infection of *Spodoptera frugiperda* insect cells with AcNPV, cytoplasmic RNA was isolated and hybrid selected on the cloned *Eco*RI fragments of AcNPV DNA (LÜBBERT et al. 1981; COCHRAN et al. 1982). The selected RNAs were in vitro translated in a reticulocyte cell free translation system (ESCHE et al. 1982; SMITH et al. 1982; LÜBBERT and DOERFLER 1984a). Proteins synthesized in vitro were labeled with radioactive amino acids and analyzed by SDS-polyacrylamide gel electrophoresis followed by fluorography. At least four early AcNPV-specific polypeptides were found which mapped in two regions of the genome (9–25 and 43–59 map units) (Fig. 1). These early mRNAs were also synthesized at late times in the infection cycle. Cell-free translation of restriction fragment-selected late AcNPV-specific RNAs (24 h postinfection) resulted in the identification and mapping of about 30 viral proteins (Fig. 2) (ESCHE et al. 1982). The translational maps in Fig. 2 were derived (A) from ESCHE et al. (1982) and LÜBBERT and DOERFLER (1984a) or (B) from SMITH et al. (1982). Both translational maps agree reasonably well. It should be noted parenthetically that the restriction maps used in the author's laboratory are based on the map of a particular isolate (TJIA et al. 1979) and differ from those of the AcNPV used in other laboratories by a single *Eco*RI site which causes deviations in nomenclature but not in the general structure of the map. For clarification, it should also be mentioned that baculo-virologists have agreed to linearize the circular AcNPV genome for illustrative purposes and to place the polyhedrin gene arbitrarily at the extreme left end of the linearized genome map (VLAK and SMITH 1982).

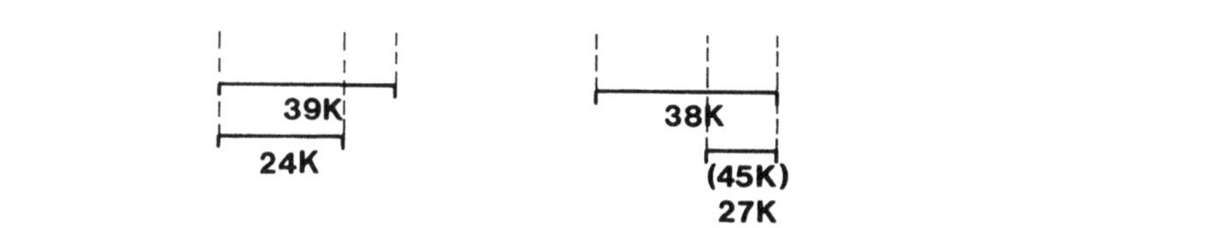

Fig. 1. The translational map of the early genes on the AcNPV genome that are expressed at about 6 h postinfection in AcNPV-infected insect cells. These functions were mapped and sized (molecular weights in kilodaltons) by hybrid selecting specific viral RNAs on cloned DNA fragments, by in vitro translating these RNAs in a reticulocyte system, and by analyzing the radioactively labeled, newly synthesized proteins by SDS-polyacrylamide gel electrophoresis (ESCHE et al. 1982). The restriction map of the AcNPV genome has been linearized and the region encoding the polyhedrin gene has been conventionally placed at the left terminus (VLAK and SMITH 1982)

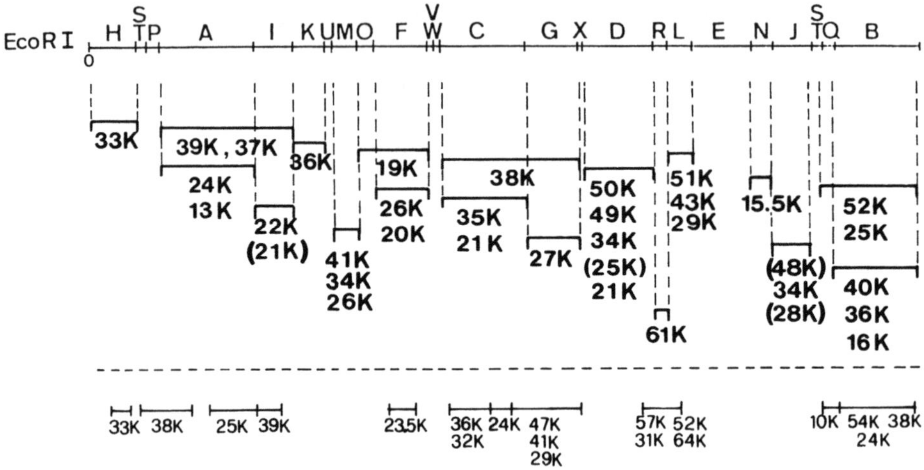

Fig. 2. The translational map of the late genes on the AcNPV genome. Details are as described in the legend to Fig. 1, except that RNA from AcNPV infected insect cells was isolated 24 h postinfection (ESCHE et al. 1982)

2.2 Temporal Regulation of the Expression of the AcNPV Genome

There is extensive evidence to indicate that from a functional point of view four classes of viral genes can be discerned with respect to the time of initiation of expression: immediate early, early, late, and very late. An immediate early gene is located between 95.0 and 97.5 map units on the linearized genome (GUARINO and SUMMERS 1986). Products of this gene are apparently required for the activation of other viral, perhaps also cellular, genes. This immediate early function may be reminiscent in its activity of the larger ElA polypeptide of adenoviruses for which such transactivating capacities have been amply demonstrated (JONES and SHENK 1979; BERK et al. 1979; NEVINS 1981). The distinction between early and late functions has most directly been drawn by the mapping of early and late viral RNAs by hybrid selection and in vitro translation (see preceding section). The early viral functions have been mapped by work in several laboratories (SMITH et al. 1982; ESCHE et al. 1982; ERLANDSON and CARSTENS 1983; LÜBBERT and DOERFLER 1984a; FRIESEN and MILLER 1985). As a corollary to this approach, an AcNPV temperature-sensitive mutant (tsB821) was isolated (MILLER et al. 1983) that was found to be defective in a very early function. At 2 h postinfection, seven AcNPV-induced proteins could be detected in mutant-infected insect cells. At the nonpermissive temperature, other viral proteins were, however, not synthesized. Viral DNA replication was inhibited or severely protracted. Host protein synthesis continued unabated in mutant-infected cells.

The enzymes involved in the transcription of the AcNPV genome have not been identified. There is evidence, however, that an AcNPV-induced or a modified host DNA-dependent RNA polymerase is responsible for the transcription of the late viral functions. This statement is essentially based upon two experimental findings (Fuchs et al. 1983): Viral transcription was fully sensitive to α-amanitin, a potent inhibitor of the cellular RNA polymerase, at early times (up to 8 h postinfection), but not at late times after infection. Moreover, a new RNA polymerase activity, which was resistant to α-amanitin and differed from the cellular enzyme, was chromatographically purified from AcNPV-infected insect cells at 24 h postinfection.

The mapping of late viral functions was accomplished in several laboratories (Smith et al. 1982; Esche et al. 1982; Vlak and van der Krol 1982; Smith et al. 1983b; Rohel et al. 1983; Lübbert and Doerfler 1984a; Rohel and Faulkner 1984; Friesen and Miller 1985). Many of the late viral RNAs appear to code for structural virion proteins, but it is by no means certain how many nonstructural AcNPV-encoded proteins exist. Moreover, next to nothing is known about cellular functions being turned on or shut off as a consequence of AcNPV infection.

From the number and size of polyhedral inclusion bodies usually observed in AcNPV-infected cells, it can readily be concluded that the polyhedrin gene must be one of the strongly expressed viral genes late in infection. It had been established several years ago that the polyhedrin constitutes an AcNPV-encoded polypeptide (Vlak et al. 1981). It had been demonstrated that late after AcNPV-infection of *Spodoptera frugiperda* insect cells, and after a cascade of new proteins had been synthesized, polyhedrin was the major polypeptide detectable (Carstens et al. 1979). In fact, late in infection a second polypeptide, a 10K protein, whose function has not yet been determined, is also produced in considerable abundance (Rohel et al. 1983; Smith et al. 1983b). Accordingly, two abundant messenger RNAs which were, of course, polyadenylated, were isolated between 30 and 48 h after AcNPV infection of *Spodoptera frugiperda* cells. By direct in vitro translation experiments, the larger one of the two messenger RNAs could be assigned to the viral polyhedrin, the smaller one to the AcNPV-encoded 10K molecular weight protein (Rohel et al. 1983; Smith et al. 1983b).

The temporal regulation of expression of the AcNPV genome appears to be quite complex (Rohel and Faulkner 1984; Friesen and Miller 1985). There is a sequential and often transient expression of several of the AcNPV-induced polypeptides. Four distinct phases of viral gene expression have been suggested to be discernible (Friesen and Miller 1985): the immediate early α-phase (2 h after infection), the early β-phase (6 h), which encompasses the onset of viral DNA replication, the late γ-phase (10–12 h), and the very late δ-phase (later than 12–15 h). During the latter phase, the production of viral inclusion bodies is initiated.

2.3 Absence of RNA Splicing

One of the striking characteristics of eukaryotic gene arrangement and expression is the splicing of RNA, a complex, enzymatically or autonomously proceeding mechanism, that eventually converts the primary transcription product into translatable messenger RNA. The mechanism of RNA splicing is also found in insect cells such as those of *Drosophila melanogaster*, although perhaps not to the same extent as in mammalian cells. It was, therefore, of considerable interest to investigate the transcripts from the AcNPV genome in insect cells for the occurrence of splicing products. So far, no evidence for RNA splicing could be adduced (LÜBBERT and DOERFLER 1984a; FRIESEN and MILLER 1985; HOWARD et al. 1986). At least 11 early and about 90 late RNAs of different size classes were mapped (LÜBBERT and DOERFLER 1984a) by conventional RNA (Northern) blot hybridization (ALWINE et al. 1977) on the AcNPV genome by using cloned restriction endonuclease fragments. When the RNA size class patterns derived from individual regions of the viral genome were compared between nuclear and cytoplasmic RNAs, significant differences could not be observed. Analyses by the Berk-Sharp method (BERK and SHARP 1977) using the Sl nuclease and by resolving the RNA-protected, labeled DNA fragments on neutral and alkaline gels in one or in two dimensions did not indicate the occurrence of spliced RNA molecules. Of course, by this method the existence of tiny intramolecular or terminal splices could not be ruled out, but appeared unlikely (LÜBBERT and DOERFLER 1984a). This type of analysis was performed for most parts of the AcNPV genome using cloned AcNPV DNA fragments for Sl protection experiments.

Individual segments of the viral genome were tested even more rigorously for the presence of spliced genes by a number of different methods. Again, to date there is no evidence for RNA splicing in this virus system. Most of the investigations on AcNPV RNA splicing have been performed in AcNPV-infected *Spodoptera frugiperda* cells. The question has, therefore, to be raised whether RNA splicing can be observed in this insect cell system at all. There is no definitive answer to this question as yet. The independence of viral gene expression of the splicing mechanism might be related to the ability of the virus to shut-off host gene expression.

2.4 Overlapping Sets of Viral RNAs in AcNPV-Infected Insect Cells

The insect baculovirus AcNPV contains a double-stranded, supercoiled circular genome of 126–129 kilobase pairs (kbp) in length. In cultured *Spodoptera frugiperda* insect cells, the virus replicates effectively, and early and late phases of viral genome expression are discernible. The large number of AcNPV transcripts and the apparent lack of splicing have raised tantalizing questions about the mechanisms involved in the expression of AcNPV DNA and its regulation (LÜBBERT and DOERFLER 1984a). It is also unknown how the widely differing numbers of RNAs and polypeptides can be correlated. For this reason, the map locations of some of the RNA size classes in three different segments of the viral genome were analyzed in detail. For this purpose a novel method has been devised which has proved useful for the analyses of transcriptional patterns in complex viral genomes (LÜBBERT and DOERFLER 1984b). The *Eco*RI fragments J, O through F, and Q (cf. Fig. 1), comprising viral DNA segments between 81.8 and 86.4, 32.6 and 41.0, and 88.2 and 89.7 map units, respectively, were investigated. Surprisingly, overlapping sets of viral RNAs of various lengths and with apparent common 3′ termini in *Eco*RI fragments J (seven size classes) and O through F (four size classes) or with apparent common

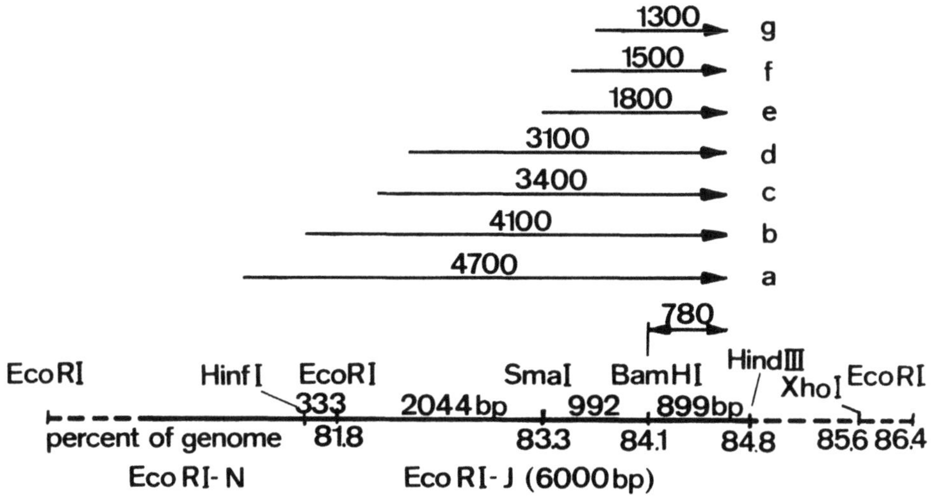

Fig. 3. Presumptive transcription pattern of AcNPV-specific RNAs in the *Eco*RI-J fragment of the viral DNA. Note overlapping sets of viral RNA with common 3′ termini. In other regions of the genome similar arrangements, sometimes with nested sets of RNAs with a common 5′ terminus, have been observed (LÜBBERT and DOERFLER 1984b; FRIESEN and MILLER 1985; RANKIN et al. 1986) It appears that the termini of these sets of RNA molecules are identical at the nucleotide level (OELLIG et al. 1987)

5′ termini in the *Eco*RI fragment Q (two size classes) were detected. The data in Fig. 3 demonstrate schematically the probable arrangement of RNA transcripts in the *Eco*RI-J segment of AcNPV DNA. At present, the functional significance of this mode of transcription is unknown. *Eco*RI fragment Q of AcNPV DNA encodes a 10K polypeptide which is abundantly expressed late after infection. The function of this protein has not yet been elucidated. The nucleotide sequences of the promoter and 5′ part of the gene for the 10K polypeptide as well as that of the entire gene were determined (KUZIO et al. 1984; LÜBBERT and DOERFLER 1984b). At least two RNAs of different lengths were transcribed in this region and initiated at one site of three nucleotides. Thus, studies on the expression of the AcNPV genome revealed interesting properties not commonly found in other eukaryotic systems.

Similar observations were made for transcripts of the AcNPV genome in the region of map units 85.1–87.5 (FRIESEN and MILLER 1985) and in the 10K gene (RANKIN et al. 1986). These authors came to the conclusion that the large number of overlapping RNA molecules might reflect a special mechanism of sequential temporal control of AcNPV gene expression by which different promoters were sequentially activated depending on the time in the infection cycle of AcNPV in insect cells. This possibility will have to be critically evaluated.

In an attempt to refine the structural analyses in the *Eco*RI-J fragment (map units 81.8 and 86.4) further, a segment comprising 4889 nucleotide pairs which included a small part of the right end of the *Eco*RI-N fragment (Figs. 1, 2), was sequenced (OELLIG et al. 1987). Several open reading frames could be distinguished which all lie on the rightward transcribed strand with respect to the conventional genome map of the AcNPV genome (Figs. 1, 2). The leftward transcribed strand did not exhibit open reading frames of appreciable lengths.

2.5 The Promoter of the Late *p10* Gene from the AcNPV Genome Can Be Inactivated by Site-Specific DNA Methylation

The methylated nucleotide 5-methyldeoxycytidine (5-mC) in specific promoter sequences of eukaryotic genes can cause the long-term inactivation of these genes (DOERFLER 1981, 1983, 1984; RIGGS and JONES 1983). The promoters of viral (KRUCZEK and DOERFLER 1983; SIMON et al. 1983; LANGNER et al. 1984, 1986; KNEBEL et al. 1985) and nonviral (BUSSLINGER et al. 1983; KESHET et al. 1985) genes were inactivated by the in vitro methylation of certain specific nucleotide sequences, e.g., of 5′-CCGG-3′ sequences. Methylations of the bodies of these genes at 5′-CCGG-3′ sequences, however, in general had no effect on gene expression (BUSSLINGER et al. 1983; KRUCZEK and DOERFLER 1983; LANGNER et al. 1984, 1986), except in the thymidine kinase gene (KESHET et al. 1985).

It is still unclear to what extent 5-mC could play a similar regulatory role in the gene expression of nonvertebrate eukaryotes, e.g., in insects. Most attempts have failed to find 5-mC in *Drosophila* DNA (SMITH et al. 1982; EICK et al. 1983) or in the DNA of *Spodoptera frugiperda* insect cells (T. MÜLLER and W. DOERFLER, unpublished results) which were used in most experiments with AcNPV.

In lepidopteran insect cells infected with the baculovirus AcNPV, two major late viral gene products are expressed: polyhedrin, a 28.9K protein, and a 10K protein (*p10*) (ROHEL et al. 1983; SMITH et al. 1983b) whose function is unknown. The nucleotide sequences of these strong promoters conform to those of other eukaryotic promoters and are rich in AT base pairs (HOOFT VAN IDDEKINGE et al. 1983; KUZIO et al. 1984; LÜBBERT and DOERFLER 1984b). The pSV0-CAT construct (GORMAN et al. 1982) containing the prokaryotic gene chloramphenicol acetyl transferase (CAT) was used to study the function of the *p10* gene promoter in insect and mammalian cells (KNEBEL et al. 1985). Upon transfection of the pAc*p10*-CAT construct, which contained 402 bp of the *p10* gene of AcNPV DNA in the *Hind*III site of pSV0-CAT, CAT activity was determined. The *p10* gene promoter was inactive in human HeLa cells and in uninfected *Spodoptera frugiperda* insect cells. The same promoter was active, however, in AcNPV-infected *Spodoptera frugiperda* cells and exhibited optimal activity when cells were transfected 18 h after infection with AcNPV. This finding demonstrated directly that the *p10* gene promoter required other viral gene products for its activity in insect cells. The nature of these products was unknown. The *p10* gene promoter sequence contained one 5′-CCGG-3′ site 40 bp upstream from the cap site of the gene and two such sites 178 and 192 bp downstream from the ATG initiation codon inside the gene. Since *Drosophila* DNA or *Spodoptera frugiperda* DNA contained no detectable 5-mC, it was interesting to determine the effect of site-specific methylations on the *p10* gene insect virus promoter. Methylation at the 5′-CCGG-3′ sites led to a block of this promoter. These data showed that an AcNPV insect virus promoter could be rendered sensitive to site-specific methylations in insect cells; albeit the DNA of these cells did not contain traceable amounts of 5-mC (KNEBEL et al. 1985). The modified nucleotide 5-mC probably constitutes a long-term signal to inactivate eukaryotic promoters. Apparently, this signal can be generally recognized as a signal inactivating gene expression, even in systems that do not seem to make extensive use of this signal.

3 The Expression of Foreign Genes in Insect Cells Through a Baculovirus Vector

3.1 The Vector System

The development of the baculovirus genome to a eukaryotic expression vector (SMITH et al. 1983a) is based mainly on the following considerations:

(1) The viral genome of some 126–129 kbp (LÜBBERT et al. 1981) can incorporate relatively large segments of additional foreign DNA.
(2) Since the virus is nonpathogenic to vertebrate cells and cannot replicate, express, or integrate its genes in vertebrate cells (TJIA et al. 1983), recombinant baculoviruses can be considered inherently safe vectors.
(3) The polyhedrin gene is a nonessential segment of the viral genome (FRASER and HINK 1982) and provides a suitable site for the insertion of foreign DNA.
(4) The expression of the polyhedrin gene is governed by a strong promoter that is active late in infection when most viral and host genes are turned off.
(5) Insertion of foreign DNA into the polyhedrin gene obviously causes deletions and functional inactivations of this gene. Hence, cells infected with recombinant virus will not contain polyhedra and could be distinguished from cells infected with wild-type virus.
(6) Finally, insect cells as eukaryotic cells are expected to be capable of accomplishing most post-translational modifications that are required to render a foreign gene product biologically active. This latter aspect is not unique to the baculovirus-insect cell system, but applies generally to eukaryotic expression vectors.

3.2 The Experimental Procedure

A number of vector constructs have been devised which all consist of a prokaryotic plasmid, the polyhedrin promoter with variable lengths of upstream sequences, an insertion site for the foreign gene, and a stretch of the polyhedrin gene proper. Details of some of the vector constructs currently used are listed in *A Manual of Methods for Baculovirus Vectors and Insect Cell Culture Procedures* by M.D. SUMMERS and G.E. SMITH (1986). Three of the AcNPV-derived

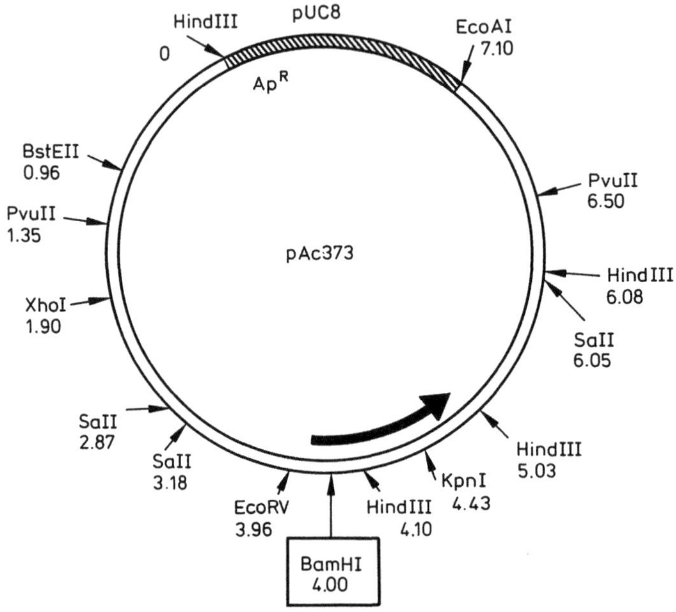

Fig. 4. Restriction map of the pAc373 vector. The unique *Bam*HI site (*boxed*) is used for the insertion of foreign DNA. This figure was taken from SUMMERS and SMITH (1986) with permission

expression vectors, pAc373 (SMITH et al. 1983a), pAc610, and pAc611 (SUMMERS and SMITH 1986) and their structures are depicted in Figs. 4 and 5.

After the foreign gene has been appropriately inserted into the vector, preferably without extensive tailing that might interfere with efficient expression, the construct is cotransfected with authentic wild-type AcNPV DNA into *Spodoptera frugiperda* insect cells. It is thought that the AcNPV promoter-foreign gene construct is inserted into the polyhedrin gene in the AcNPV genome by homologous recombination. The rate of this recombination event is apparently quite high (SMITH et al. 1983b). It is untested how and where exactly in the polyhedrin gene the insertion takes place. As a consequence of the foreign gene insertion, the polyhedrin gene is inactivated, and – as mentioned above – insect cells infected with the recombinant virus fail to produce polyhedra. Polyhedra-free plaques can therefore be recognized and isolated. This diagnostic procedure can be exploited to screen for recombinant virus. This screening method requires experience with the biological assay system. It may therefore prove advisable to complement this screening procedure by a test in which the nucleic acid from individual virus plaques is analyzed by DNA-DNA hybridization with the cloned foreign DNA as hybridization probe. The selected recombinant virus is subsequently plaque purified over several passages and propagated for more detailed analyses. Insect cell cultures infected with the recombinant virus are then investigated for production of the foreign gene product.

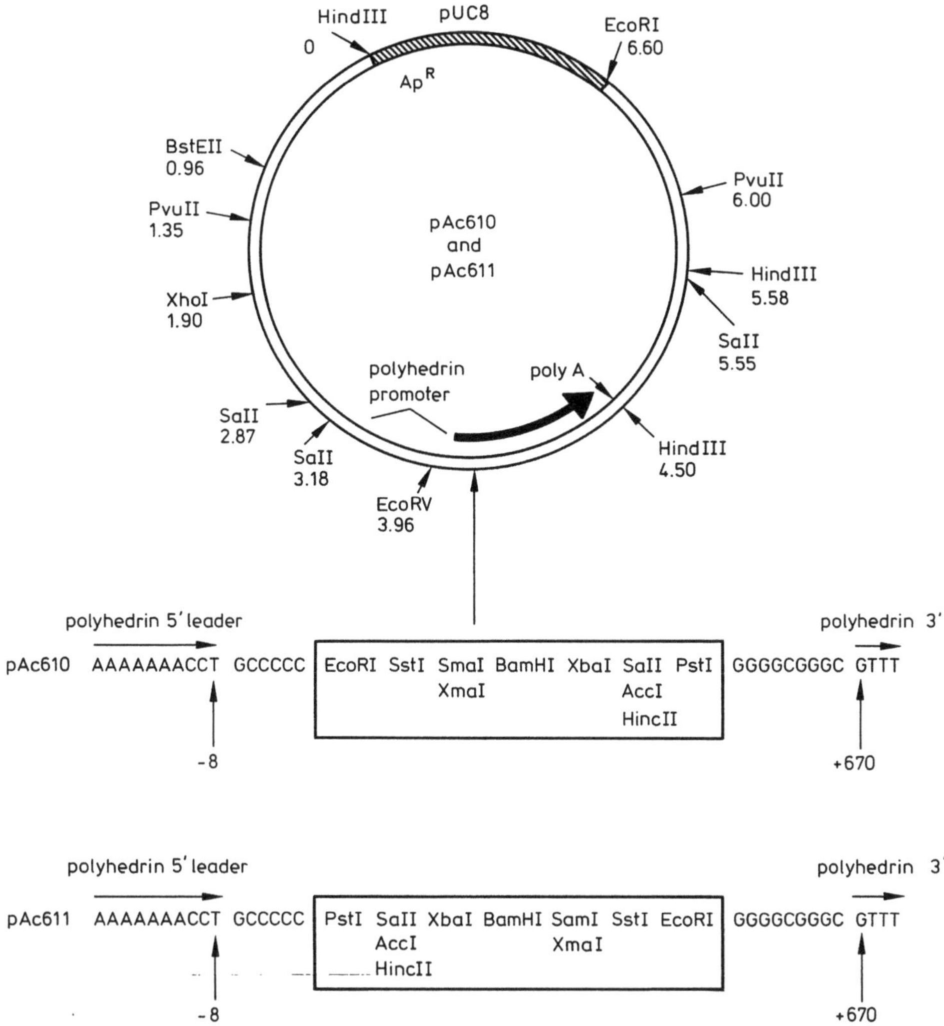

Fig. 5. Restriction maps of the pAc610 and pAc611 vectors with polylinker site constructions for the insertion of foreign DNA. This figure was taken from Summers and Smith (1986) with permission

3.3 Foreign Genes Expressed in Insect Cells by the AcNPV Vector

3.3.1 Human β-Interferon Gene (Smith et al. 1983a).

It turned out to be of importance where in relation to the polyhedrin promoter sequences the β-interferon gene was inserted. The biologically most active recombinant AcNPV constructs produced β-interferon in large amounts in infected insect cells, and >95% of the β-interferon was secreted into the maintenance medium. In 10^6 cells, levels of up to 5×10^6 units of β-interferon activity were synthesized. In bacterial and vertebrate host vectors, β-interferon levels lay considerably below this value. There was evidence that β-interferon was glycosylated, and that the signal peptide at the N-terminus was cleaved off post-translationally to produce biologically active interferon in insect cells.

3.3.2 Human α-Interferon Gene (MAEDA et al. 1985)

This gene was inserted into the baculovirus genome of *Bombyx mori* nuclear polyhedrosis virus (BmNPV), a molecule of >100 kbp, by techniques similar to those outlined above. Stable recombinant BmNPV was isolated and could be shown to replicate in silkworm larvae. These larvae contained up to 5×10^7 units of α-interferon in their hemolymph, an amount equivalent to about 50 μg. It is thought that silkworm hemolymph is characterized by the presence of potent inhibitors of proteolytic enzymes. Since the propagation of silkworms has been developed to a highly efficient industry operating at low cost, it can be expected that this expression vector system will be commercially developed to its fullest potential.

3.3.3 Escherichia coli β-Galactosidase Gene (PENNOCK et al. 1984)

This well-studied prokaryotic gene was fused to the polyhedrin promoter and recombined into the AcNPV genome. Positive virus plaques could be identified as "blue plaques" by using the β-galactosidase indicator, 5-bromo-4-chloro-3-indolyl-β-D-galactopyranoside. In these studies, it was particularly interesting to note that the regulation of the inserted prokaryotic gene came under the same temporal control that the polyhedrin promoter is subject to in that expression began late (20 h) after infection and reached a high level by 48 h postinfection.

3.3.4 Escherichia coli Chloramphenicol Acetyltransferase (CAT) Gene (CARBONELL et al. 1985)

The CAT gene was fused to the promoter from the long terminal repeat in Rous sarcoma virus and inserted into the AcNPV genome as described above. CAT activity could allegedly be demonstrated in cell lines from *Drosophila melanogaster* and *Aedes aegypti*. Hence, the retroviral promoter was active in dipteran cell lines. Baculovirus DNA was capable of a few rounds of DNA replication in *Drosophila* cells.

3.3.5 Human c-myc Protein Gene (MIYAMOTO et al. 1985)

Again under the control of the strong polyhedrin promoter, the gene (cDNA) for the human *c-myc* protein was expressed in *Spodoptera frugiperda* insect cells. This protein was characterized by immunoprecipitation and was found predominantly in the intranuclear compartment. The *c-myc* protein proved to be phosphorylated.

3.3.6 Human Interleukin 2 Gene (SMITH et al. 1985).

Infection of insect cells with the AcNPV recombinant carrying the cDNA for human interleukin 2 under polyhedrin promoter control led to the production of large amounts of the interleukin 2 polypeptide. Most of the 15.5K protein was excreted into the maintenance medium. Moreover, the recombinant interleukin 2 activated the replication of an interleukin-dependent cell line. Finally, there was evidence that the N-terminal signal peptide had been correctly cleaved off in insect cells.

3.3.7 Influenza Virus Hemagglutinin Gene (KURODA et al. 1986; POSSEE 1986)

A cDNA sequence of the influenza (fowl plague) virus hemagglutinin gene was inserted into the *Bam*HI site of the pAc373 polyhedrin vector. *Spodoptera frugiperda* cells were cotransfected with this construct, pAc-HA651, and authentic AcNPV DNA. Recombinant virus was selected by adsorption of transfected cells to erythrocytes followed by serial plaque passages on *Spodoptera frugiperda* cells. The site of insertion of the hemagglutinin gene into the AcNPV genome was determined by restriction enzyme cleavage and Southern blot hybridization analyses using hemagglutinin cDNA as probe. The influenza hemagglutinin gene was located in the polyhedrin gene of AcNPV DNA. Immunofluorescent labeling, immunoprecipitation, and immunoblot analyses employing specific antisera revealed that *Spodoptera frugiperda* cells produced immune reactive hemagglutinin after infection with the recombinant virus. The hemagglutinin was expressed at the cell surface and exhibited hemolytic capacity that had been activated by post-translational proteolytic cleavage. When chickens were immunized with *Spodoptera frugiperda* cells expressing hemagglutinin, the animals developed hemagglutination-inhibiting and neutralizing antibodies and were protected from infection with fowl plague virus. These observations demonstrated that the hemagglutinin was processed in insect cells in a similar fashion to that in fowl plague virus-infected vertebrate cells, and that the hemagglutinin had full biological activity (KURODA et al. 1986). There is evidence that the hemagglutinin is glycosylated and that it can also be expressed in insect larvae. In similar experiments, influenza virus hemagglutinin was also expressed by POSSEE (1986). The influenza virus polypeptides involved in the formation of a polymerase complex were also synthesized with the aid of the baculovirus vector system (R.M. KRUG, personal communication). Yet another viral system of which capsid proteins were synthesized with the aid of the baculovirus vector system was rotaviruses (M.K. ESTES, personal communication). Moreover, gene products of the lymphocytic choriomeningitis arenavirus were also expressed in *Spodoptera frugiperda* cells (MATSUURA et al. 1986).

These examples of foreign gene expression in insect cells demonstrate the usefulness of the baculovirus system as a eukaryotic expression vector. It can be assumed that the list of genes investigated in the studies summarized here will be outdated once this chapter has appeared in print. The baculovirus insect cell system presents a highly efficient expression machinery for prokaryotic and eukaryotic genes. I also wish to argue that this vector system is one of the safe eukaryotic virus vectors because of the demonstrated inability of AcNPV to replicate, express, or integrate its genes under the conditions of baculovirus infection in mammalian cells (Tjia et al. 1983). With the demonstration that larvae of *Bombyx mori* can be manipulated to produce gene products of biomedical importance in large amounts and to carry them in the hemolymph, this vector system will probably be scrutinized for large-scale industrial production. However, one of the still unresolved questions is the feasibility of foreign gene product purification from this biological source.

4 Perspectives

There are several major tasks to be accomplished before one can hope to understand the mechanisms underlying the expression of the baculovirus genome. There is still a lot of work to be completed to recognize and map all the functions encoded in this genome. The most direct approach to this complex problem will be the determination of the nucleotide sequence of the entire viral genome, which is not a minor endeavor due to the size of the genome. A complementary possibility might be to express certain parts of the viral genome under the control of the strong polyhedrin promoter in insect cells with the aid of the baculovirus vector system described.

It is apparent from the first survey type of studies on baculovirus gene expression that there is temporal regulation of gene activity. However, the factors affecting this regulated expression are totally unknown. Similarly, the contributions of host factors to the expression of the viral DNA are enigmatic. It might be interesting to study in considerably more detail the homologous regions (*hr*) of AcNPV DNA that have been detected at distinct sites in the viral genome (Cochran and Faulkner 1983; Arif and Doerfler 1984). The nucleotide sequences of the five homologous regions of AcNPV DNA have been determined. The consensus sequences in each region exhibit 88% homology (Guarino et al. 1986). There is evidence that the *hr* sequences have enhancer activity (Guarino et al. 1986) in that it has been shown that *hr* sequences can enhance the expression of a delayed early gene (Guarino and Summers 1986).

Another aspect in need of more research is the mechanism of viral DNA replication and the link between DNA replication and expression control, particularly with respect to the regulation of the switch from early to late genes.

From a functional point of view, almost the entire genome of AcNPV is *terra incognita*. Neither the early nor the late viral gene products, although mapped and sized by the standard and readily applied methods of molecular biology, have been biochemically and functionally characterized. Even the assignment of the structural proteins of the virion to polypeptides on the translational map is incomplete.

Promising inroads have been made in the identification of a virus-induced DNA-dependent RNA polymerase, although it is still unknown whether this enzyme is actually virus encoded. Are there additional AcNPV-induced viral factors that are required for the expression of the baculovirus genome and for its controls?

Apart from the advantages that baculoviruses offer as tools to study the molecular biology of insect cells, these viruses have attracted general interest as pliable expression vectors of eukaryotic and prokaryotic genes in insect cells. The very strong promoter of the polyhedrin gene has assured efficient expression of most polypeptides tested. These polypeptides have been post-translationally processed and modified and have been converted to biologically active products. So far, the baculovirus vector system is based on transient expression, with the infected cells eventually being killed. It will be an aim worth pursuing to fix some of the polyhedrin promoter-based constructs in the genome of the transfected cells and to achieve more permanent expression. This goal will not be easily attainable.

Apart from the obvious practical applicability, this vector system will also be useful for studies on the factors and conditions that influence the activity of the polyhedrin promoter or other viral promoters that can be incorporated into expression vector constructs. The *p10* gene promoter has been used in this context in studies on the role of DNA methylation in insect cells. In this sense, investigations on regulatory functions affecting the activity of viral promoters have just begun. Of particular interest was work on the immediate early viral gene(s) with the capacity to transactivate other viral and perhaps cellular genes as well.

Acknowledgments. I am indebted to Petra Böhm for excellent editorial work. Research in the author's laboratory was supported over successive periods by the Bundesministerium für Forschung und Technologie, Bonn-Bad Godesberg (grants number PTB 03 8497 B4 and PTB 03 8498 2) and by the Deutsche Forschungsgemeinschaft, Bonn-Bad Godesberg, through SFB74-C1.

References

Adang MJ, Miller LK (1982) Molecular cloning of DNA complementary to mRNA of the baculovirus *Autographa californica* nuclear polyhedrosis virus: Location and gene products of RNA transcripts found late in infection. J Virol 44:782–793

Alwine JC, Kemp DJ, Stark GR (1977) Method for detection of specific RNAs in agarose gels by transfer to diazobenzyloxymethyl-paper and hybridization with DNA probes. Proc Natl Acad Sci USA 74:5350–5354

Arif BM, Doerfler W (1984) Identification and localization of reiterated sequences in the *Choristoneura fumiferana* MNPV genome. EMBO J 3:525–529

Berk AJ, Sharp PA (1977) Sizing and mapping of early adenovirus mRNAs by gel electrophoresis of Sl endonuclease-digested hybrids. Cell 12:721–732

Berk AJ, Lee F, Harrison T, Williams J, Sharp PA (1979) Pre-early adenovirus 5 gene product regulates synthesis of early viral mRNAs. Cell 17:935–944

Busslinger M, Hurst J, Flavell RA (1983) DNA methylation and the regulation of globin gene expression. Cell 34:197–206

Carbonell LF, Klowden MJ, Miller LK (1985) Baculovirus-mediated expression of bacterial genes in dipteran and mammalian cells. J Virol 56:153–160

Carstens EB, Tjia ST, Doerfler W (1979) Infection of *Spodoptera frugiperda* cells with *Autographa californica* nuclear polyhedrosis virus. I. Synthesis of intracellular proteins after virus infection. Virology 99:386–398

Cochran MA, Faulkner P (1983) Location of homologous DNA sequences interspersed at five regions in the baculovirus AcMNPV genome. J Virol 45:961–970

Cochran MA, Carstens EB, Eaton BT, Faulkner P (1982) Molecular cloning and physical mapping of restriction endonuclease fragments of *Autographa californica* nuclear polyhedrosis virus DNA. J Virol 41:940–946

Doerfler W (1981) DNA methylation – A regulatory signal in eukaryotic gene expression. J Gen Virol 57:1–20

Doerfler W (1983) DNA methylation and gene activity. Ann Rev Biochem 52:93–124

Doerfler W (1984) DNA methylation: role in viral transformation and persistence. Adv Viral Oncol 4:217–247

Eick D, Fritz H-J, Doerfler W (1983) Quantitative determination of 5-methylcytosine in DNA by reverse-phase high-performance liquid chromatography. Anal Biochem 135:165–171

Erlandson MA, Carstens EB (1983) Mapping early transcription products of *Autographa californica* nuclear polyhedrosis virus. Virology 126:398–402

Esche H, Lübbert H, Siegmann B, Doerfler W (1982) The translational map of the *Autographa californica* nuclear polyhedrosis virus (AcNPV) genome. EMBO J 1:1629–1633

Fraser MJ, Hink WF (1982) The isolation and characterization of the MP and FP plaque variants of *Galleria mellonella* nuclear polyhedrosis virus. Virology 117:366–378

Friesen PD, Miller LK (1985) Temporal regulation of baculovirus RNA: Overlapping early and late transcripts. J Virol 54:392–400

Fuchs LY, Woods MS, Weaver RF (1983) Viral transcription during *Autographa californica* nuclear polyhedrosis virus infection: a novel RNA polymerase induced in infected *Spodoptera frugiperda* cells. J Virol 48:641–646

Gorman CM, Moffat LF, Howard BH (1982) Recombinant genomes which express chloramphenicol acetyltransferase in mammalian cells. Mol Cell Biol 2:1044–1051

Guarino LA, Summers MD (1986) Interspersed homologous DNA of *Autographa californica* nuclear polyhedrosis virus enhances delayed early gene expression. J. Virol 60:215–223

Guarino LA, Gonzalez MA, Summers MD (1986) The complete sequence and enhancer function of the homologous DNA regions of *Autographa californica* nuclear polyhedrosis virus. J Virol 60:224–229

Hooft van Iddekinge BJL, Smith GE, Summers MD (1983) Nucleotide sequence of the polyhedrin gene of *Autographa californica* nuclear polyhedrosis virus. Virology 131:561–565

Howard SC, Ayres MD, Possee RD (1986) Mapping the 5′ and 3′ ends of *Autographa californica* nuclear polyhedrosis virus polyhedrin mRNA. Virus Research 5:109–119

Jones N, Shenk T (1979) An adenovirus type 5 early gene product function which regulates expression of other early viral genes. Proc Natl Acad Sci USA 76:3665–3669

Keshet I, Yisraeli J, Cedar H (1985) Effect of regional DNA methylation on gene expression. Proc Natl Acad Sci USA 82:2560–2564

Knebel D, Lübbert H, Doerfler W (1985) The promoter of the late *p10* gene in the insect nuclear polyhedrosis virus *Autographa californica*: activation by viral gene products and sensitivity to DNA methylation. EMBO J 4:1301–1306

Kruczek I, Doerfler W (1983) Expression of the chloramphenicol acetyltransferase gene in mammalian cells under the control of adenovirus type 12 promoters: effect of promoter methylation on gene expression. Proc Natl Acad Sci USA 80:7586–7590

Kuroda K, Hauser C, Rott R, Klenk H-D, Doerfler W (1986) Expression of the influenza virus haemagglutinin in insect cells by a baculovirus vector. EMBO J 5:1359–1365

Kuzio J, Rohel DZ, Curry CJ, Krebs A, Carstens EB, Faulkner, P (1984) Nucleotide sequence of the *p10* polypeptide gene of *Autographa californica* nuclear polyhedrosis virus. Virology 139:414–418

Langner K-D, Vardimon L, Renz D, Doerfler W (1984) DNA methylation of three 5′ C-C-G-G 3′ sites in the promoter and 5′ region inactivate the E2a gene of adenovirus type 2. Proc Natl Acad Sci USA 81:2950–2954

Langner K-D, Weyer U, Doerfler W (1986) Transeffect of the El region of adenoviruses on the expression of a prokaryotic gene in mammalian cells: resistance to 5′-CCGG-3′ methylation. Proc Natl Acad Sci USA 83:1598–1602

Lübbert H, Doerfler W (1984a) Mapping of early and late transcripts encoded by the *Autographa californica* nuclear polyhedrosis virus genome: is viral RNA spliced? J Virol 50:497–506

Lübbert H, Doerfler W (1984b) Transcription of overlapping sets of RNAs from the genome of *Autographa californica* nuclear polyhedrosis virus: a novel method for mapping RNAs. J Virol 52:255–265

Lübbert H, Kruczek I, Tjia ST, Doerfler W (1981) The cloned *Eco*RI fragments of *Autographa californica* nuclear polyhedrosis virus DNA. Gene 16:343–345

Maeda S, Kawai T, Obinata M, Fujiwara H, Horiuchi T, Saeki Y, Sato Y, Furusawa M (1985) Production of human α-interferon in silkworm using a baculovirus vector. Nature 315:592–594

Matsuura Y, Possee RD, Bishop DHL (1986) Expression of the S-coded genes of lymphocytic choriomeningitis arenavirus using a baculovirus vector. J Gen Virol 67:1515–1529

Miller LK, Trimarchi RE, Browne D, Pennock GD (1983) A temperature-sensitive mutant of the baculovirus *Autographa californica* nuclear polyhedrosis virus defective in an early function required for further gene expression. Virology 126:376–380

Miyamoto C, Smith GE, Farrell-Towt J, Chizzonite R, Summers MD, Ju G (1985) Production of human *c-myc* protein in insect cells infected with a baculovirus expression vector. Mol Cell Biol 5:2860–2865

Nevins JR (1981) Mechanism of activation of early viral transcription by the adenovirus ElA gene product. Cell 26:213–220

Newmark P (1986) Approval for first British virus release experiment. Nature 320:2

Oellig C, Happ B, Müller T, Doerfler W (1987) Overlapping sets of viral RNAs reflect array of polypeptides in the *Eco*RI-J and -N fragments (map positions 81.2 to 85.0) of the *Autographa californica* nuclear polyhedrosis virus genome. (Submitted)

Pennock GD, Shoemaker C, Miller LK (1984) Strong and regulated expression of *Escherichia coli* β-galactosidase in insect cells with a baculovirus vector. Mol Cell Biol 4:399–406

Possee RD (1986) Cell-surface expression of influenza virus haemagglutinin in insect cells using a baculovirus vector. Virus Research 5:43-59

Rankin C, Ladin BF, Weaver RS (1986) Physical mapping of temporally regulated, overlapping transcripts in the region of the 10K protein gene in AcNPV. J Virol 57:18–27

Riggs AD, Jones PA (1983) 5-Methylcytosine, gene regulation, and cancer. Adv Cancer Res 40:1–30

Rohel DZ, Faulkner P (1984) Time course analysis and mapping of *Autographa californica* nuclear polyhedrosis virus transcripts. J Virol 50:739–747

Rohel DZ, Cochran MA, Faulkner P (1983) Characterization of two abundant mRNAs of *Autographa californica* nuclear polyhedrosis virus present late in infection. Virology 124:357–365

Simon D, Stuhlmann H, Jähner D, Wagner H, Werner E, Jaenisch R (1983) Retrovirus genomes methylated by mammalian but not bacterial methylase are non-infectious. Nature 304:275–277

Smith GE, Vlak JM, Summers MD (1982) In vitro translation of *Autographa californica* nuclear polyhedrosis virus early and late mRNAs. J Virol 44:199–208

Smith GE, Summers MD, Fraser MJ (1983a) Production of human beta interferon in insect cells infected with a baculovirus expression vector. Mol Cell Biol 3:2156–2165

Smith GE, Vlak JM, Summers MD (1983b) Physical analysis of *Autographa californica* nuclear polyhedrosis virus transcripts for polyhedrin and 10,000-molecular-weight protein. J Virol 45:215–225

Smith GE, Ju G, Ericson BL, Moschera J, Lahm H-W, Chizzonite R, Summers MD (1985) Modification and secretion of human interleukin 2 produced in insect cells by a baculovirus expression vector. Proc Natl Acad Sci USA 82:8404-8408

Smith SS, Thomas CA Jr (1981) Two dimensional restriction analysis of *Drosophila* DNAs: males and females. Gene 13:395–408

Summers MD, Smith GE (1986) A manual of methods for baculovirus vectors and insect cell culture procedures.

Tjia ST, Carstens EB, Doerfler W (1979) Infection of *Spodoptera frugiperda* cells with *Autographa californica* nuclear polyhedrosis virus. II. The viral DNA and the kinetics of its replication. Virology 99:399–409

Tjia ST, Meyer zu Altenschildesche G, Doerfler W (1983) *Autographa californica* nuclear polyhedrosis virus (AcNPV) DNA does not persist in mass cultures of mammalian cells. Virology 125:107–117

Vlak JM, Krol S van der (1982) Transcription of the *Autographa californica* nuclear polyhedrosis virus genome: location of late cytoplasmic mRNA. Virology 123:222–228

Vlak JM, Smith GE (1982) Orientation of the genome of *Autographa californica* nuclear polyhedrosis virus: a proposal. J Virol 41:1118–1121

Vlak JM, Smith GE, Summers MD (1981) Hybridization selection and in vitro translation of *Autographa californica* nuclear polyhedrosis virus mRNA. J Virol 40:762–771

The Biochemistry and Molecular Biology of the Granulosis Virus That Infects *Plodia interpunctella*

R.A. CONSIGLI[1], D.L. RUSSELL[1], and M.E. WILSON[2]

1 Introduction

Like most other animals, insects are susceptible to diseases caused by viruses. A variety of viruses differing markedly in morphological, biophysical, and biochemical characteristics infect insects. Of the approximately 650 viruses that have been isolated from insects. 83% are from lepidopterans (moths and butterflies), about 14% are from hymenopterans (sawflies), and 3% are from orthopterans (grasshoppers and cockroaches), coleopterans (beetles), and dipterans (flies and mosquitoes) (IGNOFFO 1979). Currently, eight families of insect viruses are recognized (Reoviridae, Rhabdoviridae, Togaviridae, Parvoviridae, Picorna-

[1] Division of Biology, Section of Virology and Oncology, Kansas State University, Manhattan, KS 66506, USA
[2] Department of Bacteriology and Biochemistry, College of Agriculture, University of Idaho, Moscow, ID 83843, USA

Current Topics in Microbiology and Immunology, Vol. 131

viridae, Poxviridae, Iridoviridae, Baculoviridae), ranging from the small icosahedral parvoviruses to the larger, more structurally complex poxviruses. They include a picornavirus related to *Nudaurelia β* virus (MORRIS et al. 1979; REINGANUM et al. 1978) and three families in which the viral particles are occluded in a proteinaceous matrix. This feature, which is characteristic of cytoplasmic polyhedrosis viruses, entomopoxviruses, and baculoviruses, is unique to invertebrate viruses. The protein matrix is synthesized and deposited around viral particles during virus replication and contributes to the stability of the viruses in the physical environment outside the host. In general, granulosis virus (GV) enveloped nucleocapsids are singly occluded, unlike other occluded viruses in which several viral particles are embedded within each protein matrix (HARRAP and PAYNE 1979; MARAMAROSH 1968; SUMMERS 1977).

Many insect viruses infect several insect species. The cytoplasmic polyhedrosis viruses and iridoviruses exhibit the widest host-range and have been isolated from a number of insect orders (DAVID 1975; SMITH 1976). Entomopoxviruses and nuclear polyhedrosis viruses (NPVs) also infect insects belonging to a variety of orders. In contrast, infection by the GVs is limited to lepidopterans (DAVID 1975), and the insect rhabdoviruses presumably infect only *Drosophila* species (PRINTZ 1973). Generally, insect viruses attack the larvae and cause symptoms ranging from acute infection and insect death to inapparent infection. For reviews of the ultrastructures, host ranges, and pathologies of the various insect virus families, see references DAVID (1975), HARRAP and PAYNE (1979), MARAMAROSH (1968; 1977), and SMITH (1976).

Several of the insect viruses closely resemble vertebrate viruses both in structure and in biochemical properties. In particular, the entomopoxviruses, cytoplasmic polyhedrosis viruses, and sigma viruses show striking similarities to viruses from vertebrate hosts (ARNOTT et al. 1968c; ASAI et al. 1972; BERGOIN and DALES 1971; GRANADOS and ROBERTS 1970; HOWATSON 1970; PRINTZ 1973; ROBERTS et al. 1973). However, significant differences have been observed in the biological activities and host specificities of these viruses. Cross-transmission of insect viruses to vertebrates has been observed in only a few cases. Densonucleosis virus, which is highly virulent in several tissues of *Galleria mellonella* (greater wax moth), replicates in mammalian L cells and rat embryo cells (KURATAK et al. 1969; 1970). A more classic example is the susceptibility of dipteran insects to infection by the arboviruses that they transmit to vertebrate animals (SINGH 1972; WHITFIELD et al. 1973).

In contrast to other insect virus families, members of the Baculoviridae (the GVs, the NPVs, and the nonoccluded Oryctes and braconid viruses) apparently are restricted to invertebrate hosts (WILDY 1971; STOLTZ and VINSON 1979). No morphologically similar counterpart to the baculoviruses has been detected in vertebrates. This unique characteristic has generated much scientific interest in this group of viruses. In the search for alternatives to chemicals for suppression of insect populations, the use of bacteria, rickettsiae, viruses, parasitic insects, fungi, and nematodes as biological control agents is being investigated (IGNOFFO 1975). Of the insect viruses, only the baculoviruses have been recommended for field use. A limited host range is an important consideration in the safe implementation of viruses as insecticides (FAO 1973).

Interest in development of insect viruses as biological control agents has encouraged investigation of the molecular biology of GVs. Such studies are necessary to properly assess the impact of these viruses on nontarget hosts and on the environment as well as determining their effectiveness in controlling insect pests.

Fundamental research on the biochemical and biophysical properties of baculoviruses has concentrated mainly on the NPVs because several host cell systems are available for in vitro propagation of these viruses. Unfortunately, investigations of granulosis viruses at the cellular and molecular level have been limited because GVs as a group do not appear, as yet, to readily infect cells in vitro. In recent years, various studies on granulosis viruses have been undertaken to better understand the biochemical and biophysical properties of these complex viruses, with the eventual goal of utilizing this information for establishing in vitro systems to study process(es) of infection at the cellular and molecular level. The purpose of this article is to review the biochemical and biophysical properties of the granulosis viruses with emphasis on the GV that infects the Indian meal moth, *Plodia interpunctella* (PiGV). For other reviews dealing with the granulosis viruses see the articles written by CONSIGLI et al. (1983), DAVID (1978), HARRAP and PAYNE (1979), and TWEETEN et al. (1981).

1.1 Potential Application

The pathology of granulosis virus infection was first observed in larvae of the large white butterfly, *Pieris brassicae*, by PAILLOT in 1926. Some years later a similar disease was found in a cutworm, *Agrotis segetum* (PAILLOT 1934), but was not confirmed until STEINHAUS (1947) described the same pathology in the variegated cutworm, *Peridroma saucia*. The disease was given the name granulosis because infected tissue sections observed under light microscopy revealed cells packed with minute, refractile bodies or "granules." The viral character of these granules was demonstrated by BERGOLD (1948), who examined infected tissue of the pine shoot roller, *Choristoneura murinana*, by electron microscopy and found rod-shaped virus particles present in the tissue.

At the present time, over 80 species of lepidopteran insects have been found susceptible to granulosis viruses (MARTIGNONI and IWAI 1977). Many of these species are insects that infect economically important crops. The use of viruses in controlling insect populations has gained favorable attention recently due to the rapid development of insect resistance to conventional insecticides. For example, widespread resistance to chemical agents such as synergized pyrethrins or malathion, the only insecticides widely accepted for protection of stored commodities, has developed (FAO 1973). Furthermore, mounting concern regarding hazardous effects of chemical insecticides on the environment has encouraged development of integrated pest management programs that incorporate biological agents.

Insect viruses of the family Baculoviridae are likely candidates for viral insecticides for several reasons. In vivo laboratory and field studies have demonstrated the effectiveness of GV in eliminating target hosts. The GVs of *P. brassicae* and *P. rapae* are effective against their specific hosts (BILIOTTI et al. 1956; KELSEY 1957, 1958; WILSON 1960). In field studies, GVs of the cotton cutworm, cereal noctuid, and the cabbage looper have been used to control infestations and limit crop destruction (FAO 1973; ITO and SAKIYAMA 1975; ITO et al. 1977). Large-scale programs utilizing GVs have been successfully employed in the Soviet Union, Czechoslovakia, Yugoslavia, People's Republic of China, Australia, and Great Britain (ITO et al. 1975). For example, the GVs of the fir budworm, *C. muriana*, and the potato moth GV have been utilized in Canada and Australia, respectively. However, such programs have been limited in western Europe and the United States due to more stringent registration requirements (IGNOFFO 1973; EPA 1975; FRANZ 1976; KRIEG 1976; RIVERS 1976).

Of the baculoviruses isolated thus far, only the nuclear polyhedrosis viruses of *Heliothis zea* (bollworm), *Lymantria dispar* (gypsy moth), and *Orgyia pseudotsugata* (Douglas fir-tussock moth) have been approved and utilized in biological control in the United States (EPA 1975; SHIEH and BOHMFALK 1980). At the present time, the GV of the codling moth, *Cydia pomonella*, has been used in field studies and is nearing commercial application (FALCON et al. 1968; KURSTAK et al. 1970; HUBER and DICKLER 1975, 1977; JAQUES et al. 1977; BRASSEL 1978). The GV of the Indian meal moth, *P. interpunctella*, has also been investigated to determine its potential as a biological insecticide. Its effectiveness has been demonstrated for corn and wheat (McGAUGHEY 1975; KINSINGER and McGAUGHEY 1976) and for almonds and raisins (HUNTER et al. 1973; 1977; 1979).

The insecticidal potential of the granulosis viruses is just now being realized. As research progresses, there are prevailing considerations regarding overall safety, production costs, and effectiveness of these agents in integrated pest management programs. Unlike many vertebrate viruses, very little is known of the cellular and molecular biology of the granulosis viruses. In particular, there is little or no information concerning detrimental effects on nontarget hosts at the cellular and molecular level. Previous studies have attempted to assess possible effects of these viruses on whole organisms (BURGES and HUSSYE 1971; IGNOFFO 1973; BANOWETZ et al. 1976; KRIEG 1976; LAUTENSCHLAGER et al. 1977; RODER and PUNTER 1977; HUBER and KRIEG 1978). In these studies, no pathologic or toxic effects were observed in nontarget species exposed to virus. There is, however, some evidence suggesting incomplete NPV replication in a vertebrate viper cell line (McINTOSCH and MARAMOROSCH 1973) and in CHO cells (McINTOSCH and SHAMY 1980), and of persistence of viable NPV in cell cultures of human lung, leukocyte, and amnion tissues (McINTOSCH and SHAMY 1975). However, it must be emphasized that such tests have not properly evaluated effects of the virus at the cellular and molecular level. The possibilities of recombination involving viral and host genomes, latency and integration, and toxicity of various viral components have not been investigated.

Under the present guidelines proposed by the Environmental Protection Agency, the safety of disseminating GVs into the environment with consequent exposure of man, animals, and plants to exogenous agents cannot be fully investigated until each virus can be characterized (FAO 1973). Unequivocal identification of these viruses as well as assessing possible interactions with nontarget species both in vivo and in vitro is dependent upon prior characterization of the viral nucleic acid, structural and nonstructural proteins, and other components. Characterization of GVs is also necessary to:

(1) enable mutants to be detected which may arise during development and production as insecticides and which may possess increased virulence or alterations in host range;
(2) allow development of assays for detecting viral residues;
(3) permit monitoring of insect resistance to viral strains.

Basic research on the NPVs and GVs has been fostered by their potential use as insecticidal agents, particularly against insects resistant to chemicals. Pests of stored grain represent one such group of insects which are difficult to control by chemical means. The Indian meal moth is a stored grain insect which cannot be controlled by approved chemicals such as malathion, and is economically important to grain-producing states (Zettler et al. 1973). Thus, it is of great interest to develop alternative means of controlling this pest without damaging the stored product. Arnott and Smith (1968a) described a GV of *P. interpunctella* that satisfies the criteria of an effective biological insecticide. The virus is highly virulent and specific for *P. interpunctella*, and in laboratory studies effectively controlled insect populations in stored nuts (Hunter et al. 1973), and surface layers of corn and wheat (McGaughey, 1975b). Kinsinger and McGaughey (1976) demonstrated the PiGV retains its biological activity under normal field conditions, and that effective doses of the virus are low and are compatible with several chemical agents (McGaughey 1975a, b). These studies indicate that PiGV is effective as a biological control agent implemented in the field. However, knowledge of the physical and biological characteristics of PiGV was, until recently, severely lacking. Although studies of the in vivo infection process of PiGV have been reported (Arnott and Smith 1968a; 1968b; Hunter and Hoffman 1972), its effect on nontarget hosts (invertebrate, vertebrate, and plants) has not been investigated.

1.2 Gross Pathology and In vivo Replication

Several days after ingestion of granulosis virus occlusion bodies, lepidopteran larvae begin to display unusual characteristics including sluggishness and loss of appetite as the initial gross signs of viral infection (Hunter 1970). These symptoms are followed by an alteration in the color and opacity of the larvae, from the light brown, semilucent, normal state to an opaque, white, highly infected state (Hunter 1970). This opaque "whiteness" is due to massive quantities of occlusion bodies in the larval fat body (Hunter 1970).

The majority of information concerning the in vivo replication of the GVs and NPVs has been obtained from electron microscopic studies. The predominant organ of infection is the fat body which is the site of intermediary metabolism in the insect and the location where fat, proteins, and glycogen are stored

(ARNOTT and SMITH 1968a; ASAYAMA and OSAKI 1970; CARNER and BARNETT 1975; HAMM and PASCHKE 1963; HUGER 1963). Other tissues susceptible to viral replication include epidermis (BEEGLE 1979; DAVID 1978; KAWANISHI et al. 1972b), hemocytes (STAIRS et al. 1966), tracheal matrix cells (BEEGLE 1979; HUNTER et al. 1975), and malpighian tubules (HUNTER et al. 1975; SATO et al. 1977; STAIRS et al. 1966). Infection begins in the midgut as the virus is ingested and proceeds in two stages (SUMMERS 1969, 1971; TANADA and LEUTENEGGER 1970). The first stage occurs in the midgut cells lining the intestinal lumen. Progeny virus from this primary site of replication is released into the hemocoelom and carries the viral infection to other tissues, primarily the fat body. The second stage of replication occurs in these tissues where progeny virus in formed which either reinfects other tissue or becomes occluded and released upon lysis of the cell.

After ingestion viral infection is initiated by dissolution of the matrix protein (granulin). This process occurs because of the alkaline environment of the insect midgut (FAUST and ADAMS 1966; MARTIGNONI 1957) and may be aided by viral-associated or host alkaline proteases (EPPSTEIN et al. 1975; KOZLOV et al. 1975; MCCARTHY and LIU 1976; PRITCHETT et al. 1984; TWEETEN et al. 1978). The liberated enveloped nucleocapsids then attach to the microvilli of the midgut columnar cells. Following fusion of the viral envelope to the outer cell membrane of these cells, the nucleocapsids are released into the cytoplasm (HARRAP 1970; KAWANISHI et al. 1972b). Nucleocapsids attach to the cell nucleus at the nuclear pores 2–6 h postinfection and insert their DNA core into the nucleus while the capsid remains in the cytoplasm (SUMMERS 1971). The insertion of the DNA core may be an energetic process; however, the mechanism for intracellular release of the DNA from the nucleocapsids has not been elucidated. It has been proposed by WILSON and CONSIGLI (1985b) that GV uncoating may be initiated by a virion-associated protein kinase; this will be discussed more fully below.

Capsids are detectable within cells of the midgut 6–12 h postinfection. During the assembly process, the capsids are filled with fibrous strands by a spiral-like condensation (HUGHES 1972; KRIEG and HUGER 1969; PINNOCK and HESS 1978) followed by the incorporation of an electron-dense, viral nucleoprotein core into the empty shell (BASSEMIR et al. 1983). The mechanism of this condensation is unknown; however, it has been speculated that basic proteins (KELLY et al. 1983; TWEETEN et al. 1980a; WILSON and CONSIGLI 1985a, b) and/or polyamines (ELLIOTT and KELLY 1977; WILSON and CONSIGLI 1984) may play a role in this process. Progeny nucleocapsids can be found in the nuclei of midgut cells 12–18 h postinfection located in regions of dense, aggregated material distinct from host chromatin (HUNTER et al. 1975; SUMMERS 1971).

Nucleocapsids acquire their envelopes by two distinct mechanisms. They can utilize nuclear and other intracytoplasmic membranes which seem to proliferate during viral replication (TANADA and LEUTENEGGER 1970) or they can bud through the plasma membrane at sites modified by the insertion of peplomeric structures (ROBERTSON et al. 1974). The mechanism by which intracytoplasmic enveloped nucleocapids are released from the cell is not known. It has been proposed that they are incorporated into vacuoles which are transported to the basal membrane and released (HUNTER et al. 1975; ROBERTSON et al. 1974; SUMMERS 1969).

The second phase, infection of other body tissues, begins when these nonoccluded enveloped nucleocapsids are released from the cells of the midgut (HARRAP et al. 1977). Tracheal tissue which is adjacent to the midgut is often infected by viral progeny (BEEGLE 1979; HUNTER et al. 1975; TANADA et al. 1984) and hemocytes may also be infected by these particles. It has been proposed, however, that some hemocytes are directly infected by enveloped nucleocapsids released from granulin in the gut lumen which penetrate into the hemocoelom through intracellular spaces (TANADA and LEUTENEGGER 1970). As previously stated, fat body is the major site of secondary infection by the progeny virus (ARNOTT and SMITH 1968a, b; ASAYAMA and INAGAKI 1975; BENZ and WAGER 1971; HUNTER 1975; STAIRS et al. 1966). Infection and viral replication of the GV in this organ is different from that which occurs in the cells of the midgut epithelia in several ways (ARNOTT and SMITH 1968a, b). Progeny-enveloped nucleocapsids become occluded by the protein matrix (granulin) in the fat body. In addition, more cells are infected, the yield of progeny virus per cell is much higher, and the occluded progeny are released by cell lysis.

The events involved in adsorption of GV to fat body cells remain unclear. It is possible that the viral particles attach by the modified end of the envelope (peplomers) (ADAMS et al. 1977; HIRUMI et al. 1976; KAWAMOTO et al. 1977b; SUMMERS and VOLKMAN 1976) and penetration occurs by viropexis. Release of the nucleocapsid from the envelope has not been thoroughly studied.

One of the first signs of infection in the fat body cell is a dramatic increase in the size of the nucleus. The host cell chromatin is then redistributed toward the nuclear membrane and a network of electron-dense material resembling aggregated chromatin gathers in the nucleus. This material, termed virogenic stroma, becomes increasingly Feulgen-positive and the host chromatin degenerates. The virogenic stroma is the site of viral DNA synthesis and nucleocapsid assembly (XEROS 1956). In some lepidopteran species (*P. interpunctella* and *C. muriana*) the hypertrophied nucleus degenerates and the stroma mixes with cytoplasm where the nucleocapsids are formed (ARNOTT and SMITH 1968a, b; HUGER and KRIEG 1961). In other species (*Cydia pomonella, Trichoplusia ni, Phthorimaea operculella*) this degeneration does not occur as quickly and the nucleocapsids are formed in the nucleus before degeneration (BENZ and WAGER 1971; STAIRS et al. 1966, SUMMERS 1971).

The process of envelope acquisition in the cells of the fat body is complex and diverse. In the GVs infecting *P. interpunctella, Plutella xylostella*, and *Cadra cautella*, envelopes are formed from smooth endoplasmic reticulum that accumulates during replication. The nucleocapsids align themselves in arrays along the reticulum and are inserted into envelopes (ARNOTT and SMITH 1968a; ASAYAMA 1975; HUNTER and HOFFMAN 1970). GVs may also obtain their envelopes by budding through nuclear or cytoplasmic membranes of by de novo synthesis (BIRD 1964; PINNOCK and HESS 1978; STOLTZ et al. 1973; SUMMERS 1971).

Occlusion, the deposition of granulin on the viral envelope surface, begins at one side or one end of the virion and proceeds around the particle (ASAYAMA

1975; HUNTER and HOFFMAN 1970; PINNOCK and HESS 1978). This occurs only on enveloped nucleocapsids in the cells of the fat body (BERGOLD 1963b). Large masses of cells become packed with viral particles and rupture, liberating the virus into the hemocoelom, eventually causing death of the insect (HUGER 1963; ARNOTT and SMITH 1968a, b).

Limited biochemical data have been gathered to further augment the morphological description of infection. In *C. pomonella* a sharp rise in RNA synthesis in the cytoplasm, chromatin, and the nucleus of the fat body is noted using [^{3}H]uridine and uptake studies. This occurs prior to any cytopathological changes (BENZ and WAGER 1971). Twenty-five hours postinfection this RNA synthesis is localized to swelling nucleoli (WAGER and BENZ 1971) after which these nucleoli degenerate and the levels of RNA synthesis return to normal. The shift of host chromatin to the periphery of the nucleus occurs simultaneously with the degeneration of the nucleoli.

DNA synthesis remains unaffected for the first 24 h postinfection. [^{3}H]thymidine uptake peaked 30–40 h postinfection concurrently with nuclear hypertrophy and virogenic stroma formation. A second increase in incorporation was observed during nuclear membrane degeneration at the periphery of the virogenic stroma (WATANABE and KOBAYASHI 1970).

The level of DNA synthesis 60–70 h postinfection was 30 times the level of that in uninfected cells. At this time, the level of RNA synthesis was also elevated and continued at high levels until occlusion occurred; at which time there was a dramatic decrease in transcription (WATANABE and KOBAYASHI 1970).

MAEDA and TANADA (1983) examined GV protein synthesis in fat body explant cultures where they detected sequential synthesis of viral proteins. The structural proteins of the virion were synthesized early in infection (before 72 h) and the matrix protein was synthesized later in the course of infection (3–6 days). It has been observed in *Hyphantria cunea* (fallweb worm) that [^{3}H]tyrosine is incorporated into newly synthesized proteins associated with the virogenic stroma in fat body tissues (WATANABE and KOBAYASHI 1970).

Additional information on baculovirus pathogenesis can be found in recent reviews by CONSIGLI et al. (1983), FAULKNER (1981), GRANADOS (1980), KELLY (1982), TWEETEN et al. (1981), and WALKER et al. (1982).

2 Molecular Biology

2.1 Virus Structure

BERGOLD (1948) first investigated the structure of the granulosis viruses using electron microscopy. The GVs are generally ovocylindrical in shape (Fig. 1A, B) and average 300–500 nm in length by 120–350 nm in width (BERGOLD 1963b; HUGER 1963). Thin sectioning of virus (Fig. 1B) and viral infected tissues revealed that the GVs were structurally complex viruses having a rod-shaped nucleocapsid which was enclosed in a unit membrane envelope (ARNOTT and SMITH 1968a, b; ASAYAMA 1975; BERGOLD 1963a; MORGAN et al. 1955; STAIRS et al. 1966; SUMMERS and PASCHKE 1970). Each enveloped nucleocapsid is embedded in a protein matrix (BERGOLD 1963b). This matrix has a regular crystalline lattice pattern which is not interrupted by the enveloped nucleocapsid (BERGOLD 1963b). An electron-dense layer is observed near the periphery of this matrix (ARNOTT and SMITH 1968a; KAWANISHI et al. 1972a). Whether this surface layer is highly condensed matrix protein (HARRAP 1972a, b) or a lipoprotein membrane (HESS and FALCON 1978) has not yet been determined. The GV particles, in their embedded form, have been referred to as occluded enveloped nucleocapsids, inclusion bodies, or capsules.

BEATON and FILSHIE (1976) demonstrated that the envelope of GVs appears as three layers and is 5–6 nm in thickness. Their studies also revealed that the envelope is structurally similar to a biological lipoprotein unit membrane, and unlike other viral envelopes is not modified by glycoprotein spikes (BEATON and FILSHIE 1976).

The nucleocapsids, which consist of proteinaceous capsids and DNA-protein cores, are rod-shaped and measure 260–360 nm in length by 30–60 nm in width (TWEETEN et al. 1977a, 1981). The opposite ends of the rods are morphologically dissimilar (KOZLOV and ALEXEENKO 1967; SUMMERS and PASCHKE 1970). One end is blunt and is referred to as the tail plate while the opposite end is "capped" (TEAKLE 1969).

The capsid itself is made up of subunits which are arranged in a regular lattice (SMITH and HILLS 1962; SUMMERS and PASCHKE 1970). This lattice consists of subunits assembled in rings stacked one on top of another (BEATON and FILSHIE 1976; HARRAP and JUNIPER 1966). Recent studies indicate the presence of a nucleoprotein complex (Fig. 1F) within the capsid (TWEETEN et al. 1980a).

2.2 Isolation of Viral Components

The complex nature of the granulosis virus made it necessary to separate and purify its various structural components (protein matrix, enveloped nucleocapsid, nucleocapsid, capsid, and DNA) in order to definitively characterize the various structural proteins of the virus and its nucleic acid. Several methods of purification of the occluded virus have been described (HARRAP 1972a, b; HARRAP and LONGWORTH 1974; KHOSAKA et al. 1971; SUMMERS and EGAWA 1973; TWEETEN et al. 1977a). In general, these procedures begin with homogenization of infected larvae and include several steps of differential centrifugation and/or velocity sedimentation in gradients. TWEETEN et al. (1977a) utilized mixing experiments to demonstrate the adequacy of the purification procedure which they developed for purification of PiGV. In these experiments, uninfected, radiolabeled larvae and infected, unlabeled larvae were cohomogenized. The absence of detectable radioactivity in purified virus preparations demonstrated the effectiveness of this procedure in removing contaminating host protein. They noted the use of detergent (1% deoxycholate, v/v) is vital to isolation of pure, host protein-free, virus. This treatment does no damage to either the physical or biological integrity of the virus (Fig. 1A, B).

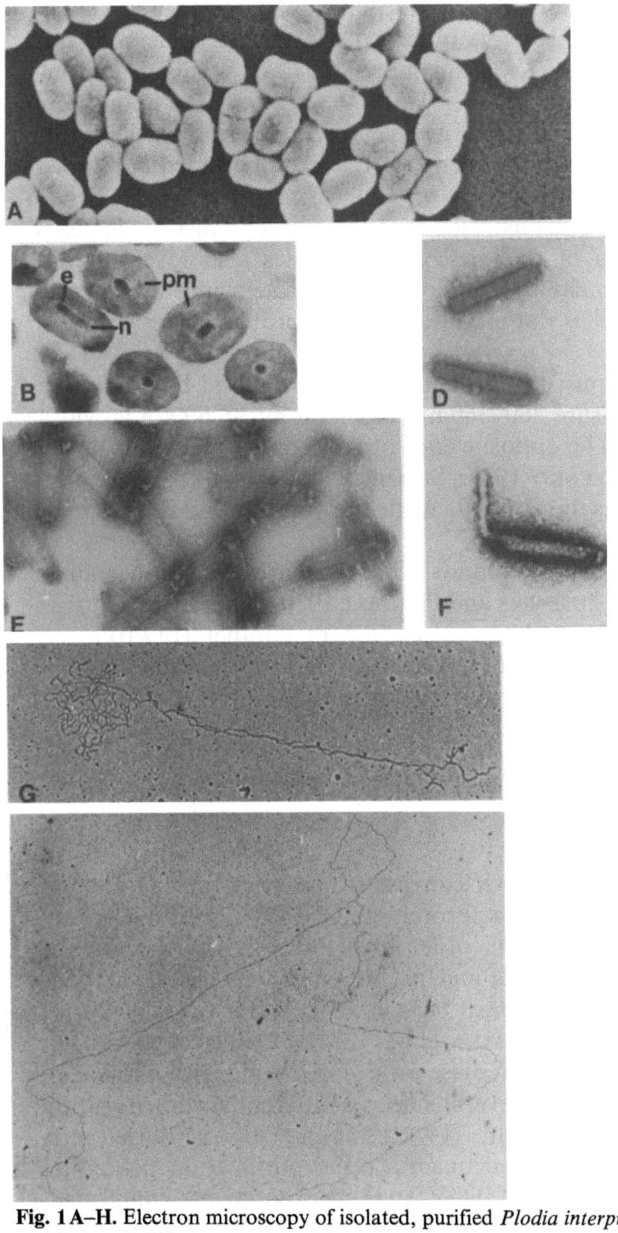

Fig. 1 A–H. Electron microscopy of isolated, purified *Plodia interpunctella* GV components. **A** Highly purified occluded virus as viewed by scanning electron microscopy. **B** Transmission electron micrograph of thin-sectioned occluded GV. Note the protein matrix (*pm*), envelopes (*e*), and nucleocapsids (*n*). **C** Enveloped nucleocapsids obtained by treatment of the occluded virus with alkaline carbonate buffer to remove the granulin. **D** Nucleocapsids obtained by treatment of the enveloped nucleocapsids with NP-40 to remove envelope. **E** Empty capsids, devoid of DNA and basic VP12, produced by treatment of nucleocapsids with a chelating agent. **F** The DNA protein complex ejecting from the capsid structure, after chelating agent treatment. **G** The DNA of the GV in its circular, supercoiled form. **H** The DNA of the GV in its relaxed circular form

Dissociation of the purified, occluded virus into enveloped nucleocapsids and matrix proteins in vitro is aided by exposure of virus to sodium carbonate buffers, pH 9–11 (BERGOLD 1947; EGAWA and SUMMERS 1972; HARRAP 1972b; KHOSAKA et al. 1971; MCCARTHY and LIU 1976; SUMMERS and PASCHKE 1970; TWEETEN et al. 1978). The conditions (i.e., pH, temperature, concentration of carbonate) for dissolution of the protein matrix must be optimized for each GV (EGAWA and SUMMERS 1972). TWEETEN et al. (1978) outline a procedure for PiGV which involves subjecting pure, occluded virus to a solution of 0.05 *M* $NaCO_3$, 0.05 *M* NaCl, pH 10.6 at room temperature for 5–15 min for complete dissolution of the granulin. Enveloped nucleocapsids are separated from the matrix proteins by velocity sedimentation on glycerol gradients (Fig. 1C). The enveloped nucleocapsids band within the gradient and the granulin remains near the top.

To release the viral nucleocapsids from the envelope, the nonionic detergent Nonidet P-40 (NP-40) appears to be the most effective reagent. TWEETEN et al. (1978) describe a procedure for the removal of viral envelopes from PiGV. It involves incubation of the enveloped nucleocapsids in 1% (v/v) NP-40 for 30 min at room temperature followed by sedimentation on glycerol gradients. The envelope material remains on top of the gradient and the nucleocapsids can be recovered as a band migrating into the gradient (Fig. 1D). It was noted by these investigators that maintaining the pH of the preparations at 8.5 using 0.01 *M* Tris-HCl significantly reduced viral aggregation which could cause the loss of nucleocapsids during centrifugation.

Isolation of viral capsids from enveloped nucleocapsids was described by SUMMERS and SMITH (1978). Treatment of the virus with a 0.1 *M* NaCl solution containing NP-40, EDTA, and Tris-HCl, pH 8.5, for 18 h at 37° C followed by velocity sedimentation on preformed gradients resulted in capsid preparations banding at 1.33 g/ml. TWEETEN et al. (1980c) modified this procedure for nucleocapsids by omitting the NP-40. The capsids prepared in this manner do not retain the cap-like structure located at one end (Fig. 1E). Whether the cap is unstable in high salt concentrations or whether it is associated with the genome is not known.

2.3 Structural Proteins

2.3.1 *Protein Matrix*

The predominant protein in GVs is the protein in the matrix, termed granulin (Fig. 1B). When the matrix is solubilized by alkaline buffers a heterogeneous mixture of polypeptides is generated. These polypeptides have been examined by gel electrophoresis and sedimentation analysis (LONGWORTH et al. 1972; SUMMERS and EGAWA 1973; TANADA and WATANABE 1971). Carboxy and amino-terminal analysis of the solubilized granulin indicated multiple species, and serological assays such as agar-gel immunodiffusion demonstrated that at least two antigenic species are generated. These results indicated a complex composition and a possible subunit structure for the matrix.

Further biochemical and biophysical characterization of the matrix was made possible by the discovery of alkaline proteases associated with these matrices (EPPSTEIN et al. 1975; KOZLOV et al. 1975; MCCARTHY and LIU 1976; TWEETEN et al. 1978). These proteases have been found to be activated by alkaline conditions similar to those used to solubilize the matrix and seem to cleave the matrix into lower molecular weight polypeptides (described more fully below). Protease inhibitors (e.g. HgCl and heat) allow solubilization and recovery of the matrix in the nondegraded form (EPPSTEIN et al. 1975; SUMMERS and SMITH 1975b; TWEETEN et al. 1978). This form of the matrix is composed of a single 12S component which, when treated with sodium dodecyl sulfate and 2-mercaptoethanol, is dissociated into one low molecular weight polypeptide (SUMMERS and SMITH 1975a, b). The molecular weights of the matrix proteins of the GVs fall within the range of 25K–30K (BROWN et al. 1977; CROIZIER and CROIZIER 1977; SUMMERS and SMITH 1975a, b, 1978; TWEETEN et al. 1978; YAMAMOTO and TANADA 1978b). These reports indicate that the 12S molecule consists of eight subunits of granulin.

Analysis of the amino acid composition demonstrated that the granulins from several different GVs are very similar. These proteins contain characteristically high levels of aspartic and glutamic acid and also have substantial amounts of valine, isoleucine, and leucine. The granulins of *Trichoplusia ni* granulosis virus (TnGV) and PiGV have been shown to be phosphorylated (SUMMERS and SMITH 1975a; TWEETEN et al. 1980c). When the peptide maps of several granulins as well as several polyhedrins (the matrix protein of NPVs) were compared, common peptides were revealed (MARUNIAK and SUMMERS 1978; SUMMERS and SMITH 1975a). Several investigators have proposed that these common domains represent regions of the protein which are responsible for its aggregative properties and thus have been evolutionarily conserved.

2.3.2 Enveloped Nucleocapsids and Envelopes

The enveloped nucleocapsids of a few GVs have been purified and analyzed for protein content by SDS-polyacrylamide gel electrophoresis (BROWN et al. 1977; SUMMERS and SMITH 1978; TWEETEN et al. 1980c; YAMAMOTO and TANADA 1979). Many structural proteins (from 12 to 20) have been identified and, though all the GVs are similar, each has a unique protein composition (Table 1). Polypeptides range in mol.wt. from 12K to 160K.

Analysis of proteins associated with the viral envelope has been accomplished in two ways; first by observations of the differences in profiles between the enveloped nucleocapsids and the nucleocapsids, and second by surface labeling the enveloped nucleocapsids with ^{125}I. When the profiles of enveloped nucleocapsids and nucleocapsids are compared, three to five proteins appear to be removed by detergent treatment (BROWN et al. 1977; TWEETEN et al. 1980c; YAMAMOTO and TANADA 1979). Surface-labeling the enveloped nucleocapsids using ^{125}I by the lactoperoxidase method (TWEETEN et al. 1980c) revealed five proteins exposed on the membrane surface. These five proteins, VP17, VP39,

Table 1. Structural polypeptides of GV-enveloped nucleocapsids[a]

S. frugiperda GV[b]	*T. ni* GV[b]	*P. brassicae* GV[c]	*P. unipuncta*[d] GV	*P. interpunctella*[e] GV
VP160				
VP95	VP90	VP91	VP97	VP97
VP89	VP85		VP89	VP88
	VP77			
VP71		VP73		
		VP68		
	VP64	VP64		
	VP63			VP63
VP58	VP57	VP58		
	VP56			
	VP52		VP52	
VP48	VP47	VP47	VP48	VP48
				VP46
				VP44[f]
VP42		VP42	VP42	VP42
VP39	VP39			VP39[f]
	VP38		VP38	VP38
VP37			VP37	
		VP34	VP34	
VP31	VP31	VP30		VP31
VP29			VP29	VP29[f]
VP26				VP26
	VP25			
VP24	VP24	VP24	VP24	
VP21.5	VP21.5			
VP20.5	VP20.5			
VP18	VP18	VP18		
VP17	VP17			VP17[f]
VP16	VP16			VP16
		VP12	VP14	VP12

[a] The numbers in the designations refer to the molecular weights ($\times 10^{-3}$) of the viral proteins, as determined by comparisons with molecular weight standards
[b] SUMMERS and SMITH (1978)
[c] BROWN et al. (1977)
[d] YAMAMATO and TANADA (1979)
[e] TWEETEN et al. (1980c)
[f] Indicate phosphorylated proteins

VP42, VP48, and VP97, correspond to the major species lost from enveloped nucleocapsids following detergent treatment.

Several polypeptides fall into a chasm. Though they are not present in the profile of nucleocapsid proteins, they are also not detectable when the enveloped nucleocapsids are surface iodinated. It is possible that these proteins, VP16, VP38, and VP88, are associated with the envelope but are not accessible at the outer surface. Also, these proteins may be part of an "intermediate layer" proposed by KAWAMOTO et al. (1977a) rather than part of the envelope.

The enveloped nucleocapsids of PiGV and *Heliothis zea* NPV (HzNPV) have been found to be capable of hemagglutinating several species of vertebrate erythrocytes (ANDERSON et al. 1981; SHAPIRO and IGNOFFO 1970). In the studies with PiGV six species of vertebrate erythrocytes were utilized including human, bovine, rabbit, chicken, mouse, and guinea pig. Though the viral-enveloped nucleocapsids could hemagglutinate the red blood cells of all of these species, those of the rabbit showed the highest degree of sensitivity. The conditions for this reaction were optimized. An acidic pH was determined to be required for agglutination even though virus was capable of binding to cells under both acidic and alkaline conditions. The hemagglutination was therefore believed to be facilitated by aggregation of the virus which occurred at low pH. The fact that baculovirus-enveloped nucleocapsids have the ability to recognize vertebrate cell receptors with the potential possibly to infect the cells of these species points out the need to examine these viruses more closely.

2.3.3 Nucleocapsids and Capsids

Several laboratories have analyzed the structural polypeptides of purified GV nucleocapsids by SDS-PAGE (BROWN et al. 1977; TWEETEN et al. 1980c; YAMAMOTO and TANADA 1979). Three to eight polypeptides ranging in mol.wt. from 12K to 73K have been identified in the GVs of *P. brassicae*, *P. interpunctella*, and *Pseudaletia unipuncta*. Although several differences exist in the minor protein species of these three viruses, there is a high degree of similarity with respect to two major polypeptides. The molecular weights of these two proteins are in the range of 31K–34K and 12K–16K. The larger of these two appears to be the major protein of the capsid structure as ascertained by SDS-PAGE analysis of purified capsids (SUMMERS and SMITH 1978; TWEETEN et al. 1980c). A protein species of approximately this same molecular weight has been identified in several other GVs and NPVs (BROWN et al. 1977; MERDAN et al. 1977; PAYNE et al. 1977), indicating these polypeptides are major elements of all baculovirus capsids.

The second, low-molecular weight protein also seems to be common to the baculoviruses (BROWN et al. 1977; CIBULSKY et al. 1977; GUELPA et al. 1977; HARRAP et al. 1977; KELLY et al. 1983; MERDAN et al. 1977; PAYNE et al. 1977; TWEETEN et al. 1980a, c). The isolation and characterization of this basic protein was first reported by TWEETEN et al. (1980a), who demonstrated the nucleocapsids of PiGV contain a highly basic polypeptide of molecular weight 12.6K. This protein is not present in purified capsids, indicating an internal location in the capsid and a possible association between this protein and the viral DNA. Further support for this association was obtained by electrophoresis of this basic protein in acid-urea polyacrylamide gels (Fig. 2). The protein has a high degree of mobility, intermediate to that of histones and protamines, with an isoelectric point (pI) of 9.8–10.0. Amino acid analysis further elucidated the highly basic nature of this protein by indicating a high arginine content. Treatment of PiGV with chelating agents (EDTA or EGTA) caused a rupture at the capped end of the capsid (Fig. 1E) and the subsequent emergence of

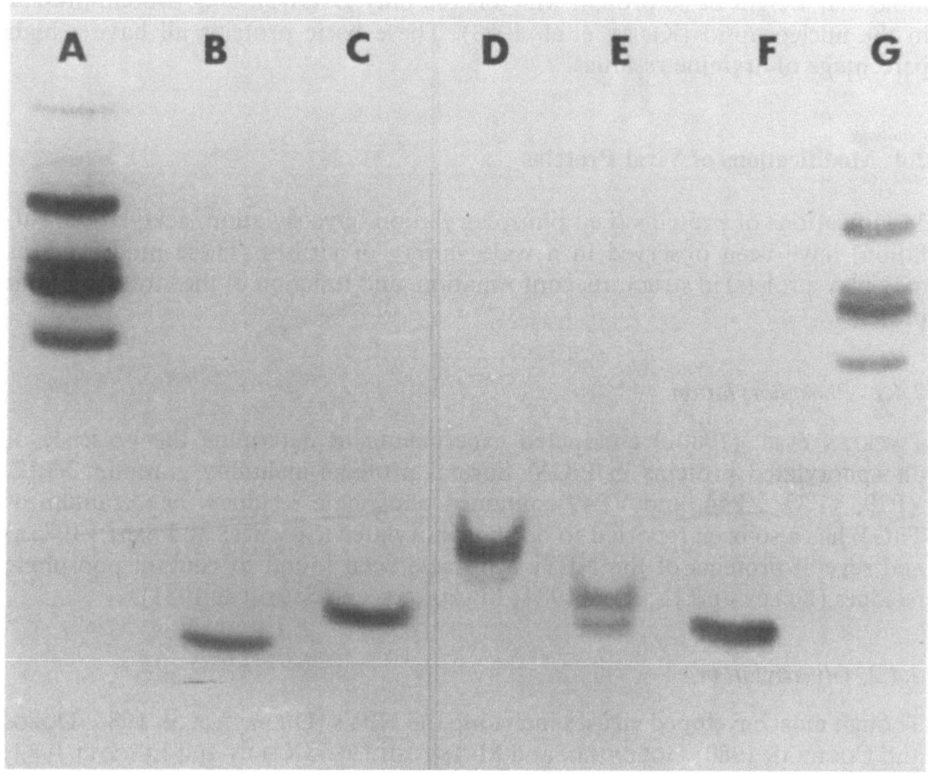

Fig. 2. Acetic acid-urea polyacrylamide gel electrophoresis of acid-soluble extracts of nucleocapsids from various baculoviruses. Acid extraction of nucleocapsids and electrophoresis were conducted as described by TWEETEN et al. (1980a). This figure is a composite of two separate gels: *A* to *C* and *D* to *G*. (*A*, *G*, calf thymus histones; *B*, *F*, acid extract from nucleocapsids of *Plodia interpunctella* GV; *C*, *D*, *E*, acid extracts from nucleocapsids of *Autographa californica* NPV, *Spodoptera frugiperda* NPV, and *Pieris rapae* GV, respectively

a thick fiber of electron-dense material (Fig. 1F). This fiber, which was sensitive to salt, was a complex of DNA and perhaps protein. Due to the sensitivity of the complex it has not yet been isolated in its intact form.

The DNA of PiGV is circular and supercoiled and has a molecular weight of 76×10^6 (TWEETEN et al. 1977b). The factors responsible for packaging the large genome into the rod-shaped capsid are not known. Unlike polyoma and SV40, the baculovirus genome is not associated with histones (BUD and KELLY 1980). It has been suggested by BUD and KELLY (1980) and TWEETEN et al. (1980a) that the extremely basic protein found in the nucleocapsid of baculoviruses is responsible for condensation of the DNA. WILSON and CONSIGLI (1985b) have shown that VP12, the basic internal core protein, is the only DNA-binding protein detectable following separation of the nucleocapsid polypeptides on SDS-PAGE. Similar findings were reported for HzNPV, *Spodoptera litura* GV, *Oryctes rhinoceros* nonoccluded baculovirus, and TnNPV; all contain a low molecular weight basic protein which is the only DNA-binding protein present in the nucleocapsid (KELLY et al. 1983). These basic proteins all have a high percentage of arginine residues.

2.4 Modifications of Viral Proteins

Modifications of proteins (i.e., phosphorylation, glycosylation, acetylation, sulfation) have been observed in a wide variety of viruses. These modifications may play a role(s) in structure, conformation, and function of the viral structural proteins.

2.4.1 Phosphorylation

TWEETEN et al. (1980c) conducted experiments to determine the presence of phosphorylated proteins in PiGV. Several proteins including granulin, VP17, VP29, VP39, VP44, and VP49 contained phosphate residues. The granulin of TnGV has also been reported to be phosphorylated (SUMMERS and SMITH 1975a) and several proteins of the NPVs also have been found to contain phosphate residues (KELLY and LESCOTT 1984; MARUNIAK and SUMMERS 1981).

2.4.2 Glycosylation

Though most enveloped viruses including the NPVs (DIABATE et al. 1983; DOBOS and COCHRAN 1980; GOLDSTEIN and MCINTOSH 1980; KELLY and LESCOTT 1983;

Table 2. Lectin binding to SDS-PAGE separated viral polypeptides

Lectin	Specific sugar	Viruses			
		Sendai[a]	NDV[a]	PiGV	AcNPV[a]
Con-A	α-D-Mannose; α-D-Galactose	75, 55	68, 52, 22	–	70, 54
DBA	α-D-N-acetyl galactosamine	75, 55	68	39, 31, 29, 26, 17, 12	70, 68, 54, 31
LPA	Sialic acid	–	–	17	–
PNA	D-Galactose-β (1–3)	–	–	–	–
RCA-1	β-D-Galactose	75, 55	68, 52	–	70, 54
SBA	D-Galactose α-D-*N*-acetyl galactosamine	75, 55	68, 52	–	–
UEA-1	α-L-Fucose	75, 55	68, 52	44, 31, 17	–
WGA	Sialic acid [β-(1–4)N-acetyl glucosamine]^2	75, 55	68, 52	–	70, 68, 54

Numbers indicate the molecular weights $\times 10^{-3}$ of the glycoproteins detected by each lectin
[a] Enveloped viruses with reported glycosylated proteins
– signifies no proteins were found to be glycosylated

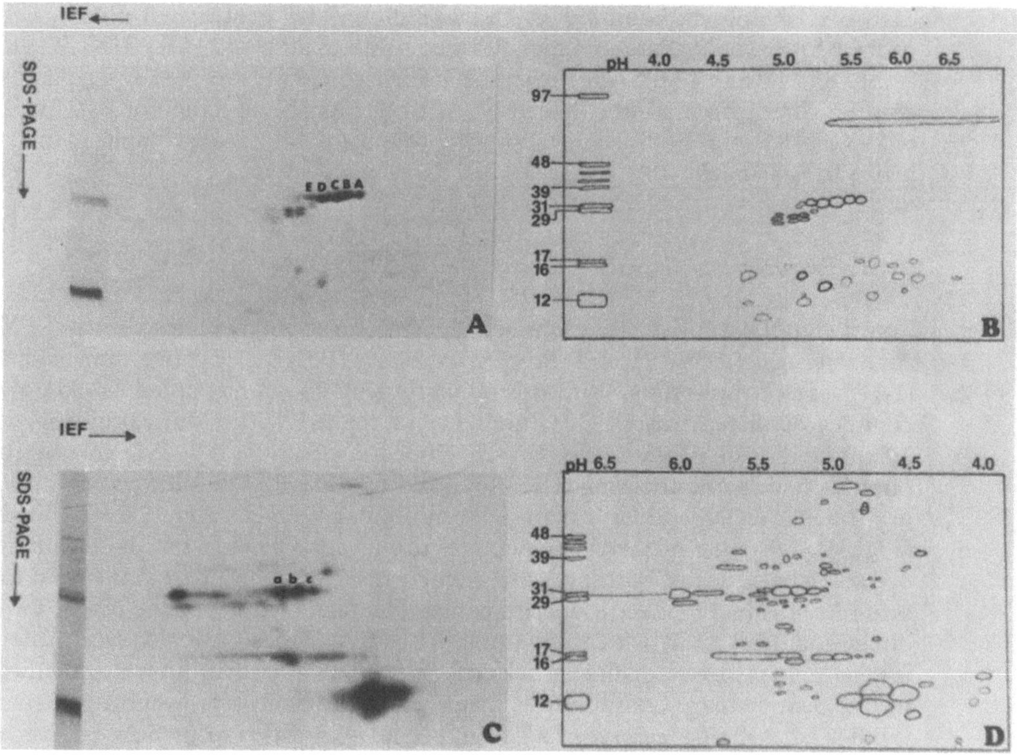

Fig. 3A–D. Two-dimensional gel analysis of the granulin and the enveloped nucleocapsids of *Plodia interpunctella* GV. Isoelectric focusing (IEF) gels were run in the first dimension followed by SDS-PAGE in the second dimension. **A** Coomassie blue stained two-dimensional gel of granulin. The five major species (*A–E*) as well as dimer (mol.wt. = 74000) and several breakdown products (mol.wt. = 27000–8000) can be seen. **B** Diagram of the gel in **A.** Molecular weights $\times 10^{-3}$ are indicated *on the ordinate* and were determined by using purified GV as a standard. The pH gradient is indicated *on the abscissa.* **C** Coomassie blue stained two-dimensional gel of purified enveloped nucleocapsids. The three major species of VP31 (the major capsid protein) are indicated (*a, b, c*). Fifty-three polypeptides are visualized on this gel. **D** Diagram of the gel in **C.** Molecular weights $\times 10^{-3}$ are indicated *on the ordinate* and were determined by using purified GV as a standard. The pH gradient is indicated *on the abscissa*

Klenk and Choppin 1970; Maruniak 1979; Stiles and Wood 1983) contain glycosylated polypeptides, glycosylation was not previously observed in the GVs. Tweeten et al. (1980c) attempted to demonstrate this modification in PiGV utilizing both the in vivo method of microinjecting radiolabeled fucose or glucosamine into infected larvae and the in vitro galactose oxidase method of labeling purified virus; however, no glycosylated polypeptides were observed. Glycoproteins have been found to be essential for infection of *T. ni* cells by *Autographa californica* NPV (AcNPV) in vitro (Stiles et al. 1983). The presence of glycosylation in PiGV was reexamined by Russell and Consigli (1985). Using various iodinated plant lectins and a sensitive in vitro lectin binding procedure in which the radiolabeled lectins were applied directly to viral proteins separated by

SDS-PAGE and immobilized on nitrocellulose, the proteins of PiGV were probed for carbohydrate moieties (Table 2). Six proteins (39.7K, 31K, 29.9K, 26.3K, 17.3K, and 12.6K) are detectable using a probe for α-D-*N*-acetyl-galactosamine. Three proteins are detectable with the probe specific for α-L-fucose (44.9K, 31K, and 17.8K) and a single protein (17.8K) is detectable with the probe for sialic acid.

2.4.3 Two-Dimensional Gel Analysis

SINGH et al. (1983) recently examined the structural polypeptides of two NPVs *A. californica* (AcNPV) and *Porthetria dispar* (PdNPV) by two-dimensional (2-D) gel electrophoresis. With this powerful technique, they noted 81–95 polypeptides in these viruses. This finding with the NPVs led our laboratory to reanalyze the polypeptides of PiGV-enveloped nucleocapsids by 2-D gel electrophoresis. Isoelectric focusing (IEF) followed by SDS-PAGE allowed resolution of 53 acidic polypeptides ranging in molecular weight from 8K to 97.3K (Fig. 3C, D). Nine of these polypeptides are glycoproteins using the technique of radiolabeled lectin blotting. The major capsid protein (31K) was separated into three distinct isoelectric species having identical tryptic peptide maps. Two-dimensional gel analysis of the granulin (Fig. 3A, B) also revealed this protein is separable into five distinct isoelectric species all of which have an identical tryptic peptide map. Granulin was identified to be a phosphoglycoprotein since it binds labeled *Ulex europaeus* agglutinin-I, a lectin which recognizes α-L-fucose moieties and can be dephosphorylated with acid phosphatase. It is believed that the various species of the VP31 and granulin proteins separated by isoelectric focusing are due to post-translational modification of these proteins (RUSSELL 1986).

2.5 Polyamines

Polyamines, naturally occurring polycationic compounds, can bind nucleic acids. This characteristic led investigators to hypothesize a possible role for polyamines in both viral multiplication and condensation of viral DNA/RNA for assembly into virion particles. ELLIOTT and KELLY (1977, 1979) reported the presence of both putrescine and spermidine in the NPVs isolated from *Spodoptera littoralis* and *H. zea*. WILSON and CONSIGLI (1984) found a polyamine-like compound associated with the nucleocapsids of PiGV. This compound is acid soluble, can be dansylated, and migrates like putrescine in ethylacetate:cyclohexane thin-layer chromatography. However, in the chloroform:triethylamine solvent system the GV polyamine-like compound does not comigrate with putrescine. This compound also displays behavior resembling a polyamine on cation exchange columns and is therefore thought to be either a modified putrescine or a new, unidentified polyamine. These investigators additionally noted the presence of several other compounds displaying characteristics of acetylated polyamines in the PiGV.

2.6 Lipids

An analysis of the lipid content of GV envelopes was conducted by YAMAMOTO and TANADA (1978a). Using the GV of *P. unipuncta* these investigators identified phosphatidylcholine and phosphatidylethanolamine as well as an unidentified phospholipid specific to virus-infected cells. Neither the neutral lipids nor glycolipids of the envelope have been reported and a comparison of larval and GV-associated lipids has not been performed. MOORE et al. (1984) examined the lipids of three NPVs and found the major phospholipid was phosphatidylcholine although large amounts of phosphatidylethanolamine were also observed. The major nonpolar lipids are sterol esters, triglycerides and free fatty acids, and sterols. A comparison between virus and insects for lipid content demonstrated that the lipid content of the virus did not reflect that of its host. Comparison of the lipid content of a specific tissue, such as fat body or midgut may, however, show a higher degree of similarity.

2.7 Viral DNA

WYATT (1952a, b) analyzed the GVs of *C. muriana* and *Choristoneura fumiferana* (spruce bud worm) in terms of their nucleic acid composition. Since bases adenine, guanine, cytosine, and thymine were present and uracil was absent, the GVs were determined to be DNA viruses. Investigation of the GVs of *Dendrolimus sibiricus* and *A. segetum* by electron microscopy revealed that the genomes of GVs are double stranded in nature as well as quite large in size (SHVEDCHIKOVA and TARASEVICH 1968). These studies also demonstrated some circular molecules of DNA, suggesting a cyclical structure for the GV genome (SHVEDCHIKOVA and TARASEVICH 1968, 1971). Analysis of the DNA of several GVs using alkaline sucrose and cesium chloride-ethidium (or propidium) bromide gradients revealed the presence of two forms of DNA (SUMMERS 1975; SUMMERS and ANDERSON 1972a, b; TWEETEN et al. 1977b). Approximately one-third of the DNA sediments in a manner characteristic of covalently closed, circular DNA (Fig. 1G) and the remainder sediments as relaxed circles (Fig. 1H). Size determinations based on these studies resulted in molecular weight calculations of more than 90×10^6. Studies of PiGV by electron microscopy confirmed the circular, supercoiled structure, and gradient sedimentation allowed the genome to be sized at a molecular weight of 76×10^6 (TWEETEN et al. 1977b). Since this time a number of GV genomes have been characterized and are of similar size (69×10^6 to 111×10^6) and conformation (BROWN et al. 1977; BURGESS 1977: TWEETEN et al. 1977b, 1980b).

Biophysical parameters, such as thermal melting point, density, and guanine+cytosine (G+C) content, have been determined for several GVs. A comparison of these data is presented in an article by HARRAP and PAYNE (1979). TWEETEN et al. (1977b, 1980b) found the following results for the genome of PiGV: the buoyant density of this genome in CsCl is 1.703 g/ml and the G+C content is 44%; between 21% and 24% of the DNA sediments as component

I (supercoiled) DNA and the length of the genome is 37 μM as calculated by electron microscopy.

The use of molecular techniques, such as restriction endonuclease cleavage and reassociation kinetics, to analyze the genomes of GVs and NPVs has provided new insights into baculovirus molecular biology. Restriction endonuclease fragmentation analysis indicates that the GVs of *S. frugiperda, T. ni, P. rapae, Estigmene acrea,* and *P. interpunctella* are not closely related at the genomic level (TWEETEN et al. 1980b; CROOK 1981; CATTANO and LANGRIDGE 1982). A comparative study of restriction fragments generated for DNAs from the *P. brassicae* and *P. rapae* GVs demonstrated 97.7% homology between the two virus DNAs, indicating these viruses are closely related (CROOK 1981). This technique has also proved useful in the analysis of the genomes of NPVs (MILLER and DAWES 1978a; ROHRMANN et al. 1978; SMITH and SUMMERS 1978, 1979), particularly in the case of related genomic variants (LEE and MILLER 1978; MILLER and DAWES 1978b). The establishment of genetic maps and relationships to phenotypic characteristics has been undertaken with a number of NPV DNAs (SMITH and SUMMERS 1980; SUMMERS et al. 1980). Unfortunately, GV DNAs have not been mapped genetically, and establishing a relationship between such a map and phenotype expression will be difficult since an in vitro culture system for these viruses does not, as yet, exist.

2.8 Viral Associated Enzymes

2.8.1 Alkaline Proteases

The initial report of an alkaline protease associated with a baculovirus occurred in 1959 (YAMAFUGI et al. 1959). Since that time alkaline proteases have been reported in many baculoviruses. These proteases are associated with the protein matrix, and under alkaline conditions, similar to those in the insect midgut, become activated. This activation in the insect midgut results in dissolution of the matrix protein and release of the enveloped nucleocapsids (SUMMERS 1971). TWEETEN et al. (1978) have purified and characterized the alkaline protease associated with PiGV. The enzyme has a molecular weight of 14K and is optimally active at a pH of 10.5 and 40° C. The observation that inactivation of the enzyme by heat treatment resulted in a decreased rate of granulin solubilization led to the proposal that the protease functions in initiation of GV infection by aiding the removal of granulin from enveloped nucleocapsids. In NPVs, it has been noted that occlusion bodies purified from larvae contain proteases whereas those purified from insect cell lines lack proteolytic activity (MARUNIAK et al. 1979; MCCARTHY and DICAPUA 1979; ZUMMER and FAULKNER 1979). Both the occluded viruses from larvae and tissue culture were able to infect insects (POTTER et al. 1978). It appears the protease, while perhaps helpful to the dissolution process, is not essential and is perhaps a host-contributed enzyme. Analysis of the digestive juices of the armyworm, *P. unipuncta*, revealed a high level of protease activity as contrasted with the hemolymph of this insect

(NAGATA and TANADA 1983). These investigators found that the protease, under low ionic conditions, is adsorbed by the capsules of the GV infecting *P. unipuncta*.

2.8.2 Protein Kinases

Protein kinase activities have been reported in both AcNPV (MILLER et al. 1983) and PiGV (WILSON and CONSIGLI 1985a, b). The enzyme activity associated with the NPV is present in both occluded virus and in tissue culture derived extracellular virus. The primary proteins phosphorylated by the enzyme in occluded virus have molecular weights of 15K and 17K with minor acceptor proteins of molecular weights 30K, 34K, 42K, and 54K. The extracellular virus proteins phosphorylated are of molecular weights 90K, 70K, 60K, 54K, 37K, 28K, and 15K. Phosvitin can act as an acceptor protein in this NPV system; however, basic proteins such as bovine serum albumin and protamine sulfate cannot. Calcium chloride, cAMP, and cGMP have no effect on the enzyme. Both serine and threonine are the target amino acids of the kinase, and Mg^{2+} was the preferred cation for the reaction.

The kinase described for the occluded form of the PiGV is associated with the viral capsid. Whether this enzyme is viral encoded or host contributed is not known, but because of its internal location it is not thought to be a host contaminant. This enzyme is also cyclic-nucleotide independent, but demonstrates a preference for the Mn^{2+} cation over the cations Mg^{2+}, Ca^{2+}, and Zn^{2+}. The predominant viral acceptor protein is the basic protein thought to be associated with the viral DNA having a molecular weight of 12.6K (VP12), and other proteins are minimally phosphorylated (24K, 26.5K, and 41.5K). Basic proteins, such as histones and protamine sulfate, can serve as acceptor proteins for the enzyme when added to the system but phosvitin cannot. These findings indicate the enzyme described by WILSON and CONSIGLI (1985a, b) is very different from that described by MILLER et al. (1983). The target amino acids in this system are serine and arginine; no threonine phosphorylation was detected, again suggesting the kinases described by these two groups are indeed different.

WILSON and CONSIGLI (1985b) propose that the kinase enzyme, because its major endogenous target is the basic VP12 DNA-binding protein, may function in both the release of the DNA from the nucleocapsid (uncoating) and the decondensation of the DNA via phosphorylation of the VP12 protein associated with the DNA.

3 In Vitro Replication

The adaptation of NPVs for replication in vitro was an important step in the study of the molecular biology of these viruses for several reasons. Intricate investigations into the growth cycle of the virus on both cellular and molecular levels were now possible. In vitro systems provided a means of quantitation of the virus by plaque assay. These systems, in conjunction with detailed genetic maps of the virus, made possible elegant studies of several viral protein genes and their promotors, and have pushed baculovirology into the forefront of genetic engineering research (Smith et al. 1983; Pennock et al. 1984).

The successful in vitro propagation of GVs has been a difficult task. Infection in primary cultures of *L. dispar* ovaries incubated with GV-infected fat bodies from *P. brassicae* was attempted but virus production was irregular and incomplete with nuclear hypertrophy and formation of occluded virus in only some of the culture vessels (Vago and Bergoin 1963). Also, replication of GV in cell lines of *Spodoptera frugiperda*, *T. ni*, and *H. zea* has been investigated (Granados 1976). In the *T. ni* cell line aberrant infection was observed. The cells contained numerous membrane structures and long cylindrical forms resembling capsids but no nucleocapsid synthesis was observed. Similar incomplete virus development has been reported after infection of alternate insect hosts with either GVs or NPVs (Hunter and Hoffman 1972) and after prolonged undiluted serial passage of NPVs in vitro (MacKinnon et al. 1974). Rubinstein et al. (1982) attempted to propagate a GV infection in primary fat body cultures. They utilized fat body from *E. acrea* and infected them with EaGV. Although EaGV morphogenesis was not observed in these cultures, an increase in specific viral proteins was detected using an Elisa assay.

Recently Naser et al. (1984) described the in vitro growth of *C. pomonella* GV in a primary embryo cell line of this insect. This group utilized infectious hemolymph collected by bleeding larvae several days postinfection and the virus was passaged by placing infectious cell-free supernates onto adherent cells. Cytopathic effects were observed 7–8 days postinfection of these cells. Electron microscopy showed cells containing occlusion bodies which harbored single rods and a protein matrix which had a regular crystalline lattice. Miltenburger et al. (1984) stresses the importance of the cellular substrate (type of cell, e.g., hemocytes, ovaries and stage of development, e.g., larval, embryo) for baculovirus replication. He further states that the cell lines most suseptible to infection by GV in vitro were of embryo origin and fairly adherent to the plastic flasks in which they were grown.

Because our laboratory felt one of the problems involved in adaptation of the granulosis viruses to tissue culture might lie in the recognition of the host cell by the envelope proteins of the occluded form of the virus, we performed some preliminary experiments using various techniques to "artifically" introduce this form of the virus into the host cell. It was hoped that once introduced into the host cell and allowed to go through one cycle of viral replication and assembly, the virus would be adapted to the host in vitro system by the host system itself.

One technique utilized to introduce the virus into the host cell involved calcium phosphate/DMSO transfection of purified PiGV DNA into *T. ni* 368 cells (Russell 1986). Using antibody directed against enveloped nucleocapsids, we could detect the synthesis of viral specific proteins in the transfected cells. However, these proteins were shown to be only of low molecular weight by immunoblot analysis of cell extracts after SDS-PAGE and transfer to nitro-

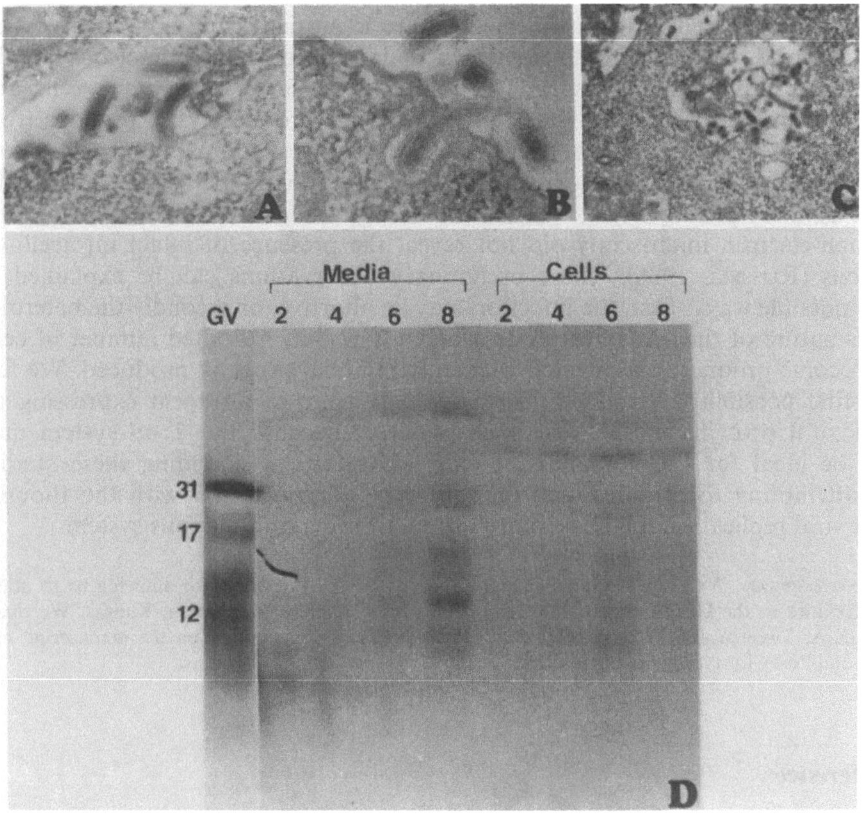

Fig. 4A–D. Attempts at in vitro replication of *Plodia interpunctella* enveloped nucleocapsids in *Trichoplusia ni* cells. **A, B** Thin-sectioned electron microscopy of enveloped nucleocapsids and *T. ni* (368) cells 30 min postfusion demonstrating GV entry into cells. **C** Thin-sectioned electron microscopy of GV-infected cells 72 h postfusion. Virus particles observed in clumps within the cells, often in vesicles. **D** Immunoblot analysis of media and cells postinfection. The sixth serial passage following an original fusion of GV with *T. ni* cells was used to infect *T. ni* cells and both media and cells were harvested separately at 2, 4, 6, and 8 days postinfection. Equal volumes of harvested material were subjected to SDS-PAGE. Purified GV enveloped nucleocapsids were also electrophoresed as a standard marker of viral proteins. The proteins were transferred electrophoretically to nitrocellulose sheets and probed with antiGV-enveloped nucleocapsid IgG and then with ^{125}I-labeled staphylococcal protein A. Increasing amounts of viral proteins were found in the media over the period of 8 days. A subsequent increase in GV proteins was observed in the cells until day 6. Day 8, however, showed a reduction in viral proteins in the cells concomitant with an increase in viral proteins in the media. Cell lysis was not observed, suggesting that a budding event may be occurring between days 6 and 8

cellulose. Because the viral genome is so large it was felt that the entire, intact nucleic acid of the virus was not being introduced into the cells. Therefore the technique of membrane fusion in the presence of polyethylene glycol was used to introduce the enveloped nucleocapsid of the virus to the host cell. Electron microscopy of the thin-sectioned fusion products indicated membrane fusion had occurred and, as a result, the entire viral genome was introduced into the cell (Fig. 4A–C). In addition to the electron microscopy studies, viral proteins in the cells and in the media were identified using immune Western blot analysis. Medium from cells which underwent enveloped nucleocapsid fusion was subsequently passaged to normal *T. ni* 368 cells and the cells and media from serial passages were probed for viral proteins. Viral proteins from the matrix were detected in cells of these passages and viral proteins from the envelope, capsid, and core were detected both in the media and in the cells of these passages (Fig. 4D); however, examination of these cells by thin section-electron microscopy did not reveal the presence of intact intracellular virions (Russell 1986). These preliminary observations can be explained in two possible ways: first, the infection may be abortive, or secondly the heterologous nature of the virus/host system may allow only a limited number of cells to become productively infected with only minimal progeny produced. We feel the first possibility is unlikely since serial passage of an agent expressing all structural proteins of the virus was possible. Because the *T. ni* system may not be ideal for PiGV replication our laboratory is extending these studies by attempting to establish cell lines from *P. interpunctella* with the thought that viral replication might be more productive in a homologous system.

Acknowledgments. We wish to extend our gratitude to Dr. Y. Pomeranz for allowing us to utilize the facilities at the USDA, Grain Marketing Research Center in Manhattan, Kansas. We thank Drs. K.A. Tweeten and W.H. McGaughey for reading and commenting on the manuscript, and Mrs. Jan Posey for typing the manuscript.

References

Adams JR, Goodwin RH, Wilcox TA (1977) Electron microscopic investigations on invasion and replication of insect baculoviruses *in vivo* and *in vitro*. Biologie Cellulaire 28:261–268

Anderson DK, Bulla LA Jr, Consigli RA (1981). Agglutination of vertebrate erythrocytes by the granulosis virus of *Plodia interpunctella*. Virology 113:242–253

Arnott HJ, Smith KM (1968a) An ultrastructural study of the development of a granulosis virus in the cell of the moth *Plodia interpunctella* (Hbn). J Ultrastruct Res 21:251–268

Arnott HJ, Smith KM (1968b) Ultrastructure and formation of abnormal capsules in a granulosis virus of the moth *Plodia interpunctella* (Hbn). J Ultrastruct Res 22:136–158

Arnott HJ, Smith KM, Fullilove SL (1968) Ultrastructure of of cytoplasmic polyhedrosis virus affecting the monarch butterfly *Danaus plexippus*. J Ultrastruct Res 24:479–507

Asai J, Kawamoto F, Kawase S (1972) On the structure of the cytoplasmic polyhedrosis virus of the silkworm, *Bombyx mori*. J Invertebr Pathol 19:279–280

Asayama T (1975) Maturation process of granulosis virus of the diamond-back moth, *Plutella xylostella*. Jpn J Appl Ent Zool 19:149–156

Asayama T, Inagaki I (1975) Cell alterations caused by the infection with the granulosis virus in the diamondback moth, *Plutella xylostella*, and the site of appearance of nucleocapsid. Jpn J Appl Ent Zool 19:79–84

Asayama T, Osaki N (1970) A granulosis virus of the diamondback moth, *Plutella xylostella*. J Invertebr Pathol 15:284–286

Banowetz GM, Fryer JL, Iwai PJ, Martignoni ME (1976) Effects of the douglas-fir tussock moth nucleopolyhedrosis virus (baculovirus) on three species of salmonid fish. Forest Service Research Paper PNW-214. US Department of Agriculture, Washington DC

Bassemir U, Miltenburger HG, David P (1983) Morphogenesis of nuclear polyhedrosis virus from *Autographa californica* in a cell line from *Mamestra brassicae* (cabbage moth). Further aspects on baculovirus assembly. Cell Tissue Res 228:587–595

Beaton SD, Filshie BK (1976) Comparative ultrastructural studies of insect granulosis and nuclear polyhedrosis viruses. J Gen Virol 31:151–161
Beegle CC (1979) Histopathology and bioassay of *Plathypena scabra* granulosis virus. J Invertebr Pathol 32:325–331
Beek CP van der, Saaijer-Riep JD, Vlak JM (1980) On the origin of the polyhedral protein of *Autographa californica* produced in cells grown in vitro. J Invertebr Pathol 22:231–237
Benz G, Wager R (1971) Autoradiographic studies on nucleic acid metabolism in granulosis-infected fat body of larvae of *Carpocapsa*. J Invertebr Pathol 18:70–80
Bergoin M, Dales S (1971) Comparative observations on poxviruses of invertebrates and vertebrates. In: Maramorosch K, Kurstak E (eds) Comparative virology. Academic, New York, pp 169–205
Bergold GH (1947) Die Isolierung des Polyeder. Virus und die Natur der Polyeder. Z Naturforsch 26:122–143
Bergold GH (1948) Über die Kapselvirus-Krankheit. Z Naturforsch 3b:338–342
Bergold GH (1963a) Fine structure of some insect viruses. J Invertebr Pathol 5:111–128
Bergold GH (1963b) The molecular structure of some insect virus inclusion bodies. J Ultrastruct Res 8:360–378
Biliotti E, Grison P, Martouret D (1956) L'utilization de'une maladie à virus comme méthode de lutte biologique contre *Pieris brassicae*. Entomophaga 1:35–44
Bird FT (1964) On the development of insect polyhedrosis and granulosis virus particles. Can J Microbiol 10:49–52
Brassel J (1978) Development of techniques for the production of a granulosis virus preparation for microbial control of the codling moth, *Laspeyresia pomonella* L (Lep Tortricidae), and estimation of production costs. Mitt Schweiz Entomol Ges 51:155–212
Brown DH, Bud HM, Kelly DC (1977) Biophysical properties of the structural components of a granulosis virus isolated from the cabbage white butterfly (*Pieris brassicae*). Virology 81:317–327
Brown M, Crawford AM, Faulkner P (1979) Genetic analysis of a baculovirus, *Autographa californica* nuclear polyhedrosis virus: I. Isolation of temperature sensitive mutants and assortment into complementation groups. J Virol 31:190–198
Bud HM, Kelly DC (1980) An electron microscope study of partially-lysed baculovirus nucleocapsid: the intranucleocapsid packaging of viral DNA. J Ultrastruct Res 73:361–368
Burand JP, Wood HA, Summers MD (1983) Defective particles from a persistent baculovirus infection in *Trichoplusia ni* tissue culture cells. J Gen Virol 64:391–398
Burges HD, Hussey NW (1971) Past achievements and future prospects. In: Burges HD, Hussey NW (eds) Microbial control of insects and mites. Academic, New York, pp 687–709
Burgess S (1977) Molecular weights of lepidopteran baculovirus DNA's: derivation by electron microscopy. J Gen Virol 37:501–510
Carner GR, Barnett OW (1975) A granulosis virus of the green clover worm. J Invertebr Pathol 25:269–271
Carstens EB, Tjia ST, Doerfler W (1979) Infection of *Spodoptera frugiperda* cells with *Autographa californica* nuclear polyhedrosis virus. I. Synthesis of intracellular proteins after virus infection. Virology 99:386–398
Cattano SP, Langridge WHR (1982) Characterization of DNA from granulosis virus of *Estigmene acrea*: (Lepidoptera) Virology 119:199–203
Ciblusky RJ, Harper JD, Gudauskas RT (1977) Biochemical comparison of virion proteins from five nuclear polyhedrosis viruses infecting plusiine larvae (Lepidoptera: Noctuidae). J Invertebr Pathol 30:303–313
Cochran MA, Carstens EB, Eaton BT, Faulkner P (1982) Molecular cloning and physical mapping of restriction endonuclease fragments of *Autographa californica* nuclear polyhedrosis virus DNA. J Virol 41:940–946
Consigli RA, Tweeten KA, Anderson DK, Bulla LA Jr (1983) Granulosis viruses, with emphasis on the GV of the Indian meal moth, *Plodia interpunctella*. Adv Virus Res 28:141–173
Crawford AM, Kalmakoff J (1977) Effect of alkaline protease on the antigenic nature of *Wiseana* nuclear polyhedrosis virus polyhedron protein. J Virol 24:412–415
Croizier G, Croizier L (1977) Evaluation du poids moleculaire de la proteine des corps d'inclusion de divers baculovirus d'insectes. Arch Virol 55:247–250
Crook NE (1981) A comparison of the granulosis viruses from *Pieris brassicae* and *Pieris rapae*. Virology 115:173–181

David WAL (1975) The status of viruses pathogenic for insects and mites. Annu Rev Entomol 20:97–117
David WAL (1978) The granulosis virus of *Pieris brassicae* (L) and its relationship with its host. Adv Virus Res 22:111–161
Diabate S, Geyer R, Stirm S (1983) Separation and sugar component analysis of the oligosaccharides in the surface glycoproteins of Newcastles disease virus. Arch Virol 76:321–334
Dobos P, Chochran MA (1980) Protein synthesis in cells infected by *Autographa californica* nuclear polyhedrosis virus (AcNPV); the effect of cytosine arabinoside. Virology 103:446–464
Egawa K, Summers MD (1972) Solubilization of *Trichoplusia ni* granulosis virus proteinic crystal. I. Kinetics J Invertebr Pathol 19:395–404
Elliott RM, Kelly DC (1977) The polyamine content of a nuclear polyhedrosis virus from *Spodoptera littoralis*. Virology 76:472–474
Elliott RM, Kelly DC (1979) Compartmentalization of the polyamines contained by a nuclear polyhedrosis virus from *Heliothis zea*. Microbiologica 2:409–413
Environmental Protection Agency (1975) Guidance for safety testing of baculoviruses. In: Summers M, Engler R, Falcon LA, Vail P (eds) Baculoviruses for insect pest control: safety considerations. American Society for Microbiology, Washington DC, pp 179–184
Eppstein DA, Thoma JA, Scott HA, Young SY, III. (1975) Degradation of matrix protein from a nuclear polyhedrosis virus of *Trichoplusia ni* by an endogenous *protease*. Virology 67:591–594
Falcon A (1976) Problems associated with the use of arthropod viruses in pest control. Annu Rev Entomol 21:205–324
Falcon LA, Kane WR, Bethell RS (1968) Preliminary evaluation of a granulosis virus of control of the codling moth. J Econ Entomol 61:1208–1213
Faulkner P (1981) Baculovirus. In: Davidson EW (ed) Pathogenesis of invertebrate microbial diseases. Allanheld, Osmun, and Co., Totowa, New Jersey, pp 5–38
Faust RM, Adams JR (1966) The silicon content of nuclear and cytoplasmic viral inclusion bodies causing polyhedrosis in lepidoptera. J Invertebr Pathol 8:526–530
Food and Agricultural Organization (1973) The use of viruses for the control of insect pests and disease vectors. Fao Agric Stud 91:1–39
Franz JM (1976) Occurrence, aspects and utilization of virus diseases of pest insects: introduction to the project. Z Angew Entomol 82:124–128
Fraser MJ, Hink WF (1982) The isolation and characterization of the MP and FP plaque variants of *Galleria mellonella* nuclear polyhedrosis virus. Virology 117:366–378
Friesen PD, Miller LK (1985) Temporal regulation of baculovirus RNA: overlapping early and late transcripts. J Virol 54:392–400
Goldstein NI, McIntosh AH (1980) Glycoproteins of nuclear polyhedrosis viruses. Arch of Virol 64:119–126
Granados RR (1976) Infection and replication of insect pathogenic viruses in tissue culture. Adv Virus Res 20:189–236
Granados RR (1978) Early events in the infection of *Heliothis zea* midgut cells by a baculovirus. Virology 90:170–174
Granados RR (1980) Infectivity and mode of action of baculoviruses. Biotechnol Bioeng 23:1377–1405
Granados RR, Roberts DW (1970) Electron microscopy of a pox-like virus infecting an invertebrate host. Virology 40:230–342
Granados RR, Nguyen T, Cato B (1978) An insect cell line persistently infected with a baculovirus-like particle. Intervirology 10:309–317
Guelpa B, Bergoin M, Croizier G (1977) La proteine d'inclusion et les proteines du virion du baculovirus du diptere *Tipula paludosa* (Meigen) CR Acad Sci 284:779–782
Hamm JJ, Paschke JD (1963) On the pathology of a granulosis virus of the cabbage looper *Trichoplusia ni* (Huber). J Insect Pathol 5:187–197
Harrap KA (1970) Cell infection by a nuclear polyhedrosis virus. Virology 42:311–318
Harrap KA (1972a) The structure of nuclear polyhedrosis viruses. I. The Inclusion body. Virology 50:114–123
Harrap KA (1972b) The structure of nuclear polyhedrosis viruses. II. The virus particle. Virology 50:124–132
Harrap KA, Juniper BE (1966) The internal structure of an insect virus. Virology 28:175–178

Harrap KA, Longworth JF (1974) An evaluation of purification methods for baculoviruses. J Invertebr Pathol 24:55–62
Harrap KA, Payne CC (1979) The structural properties and identification of insect viruses. Adv Virus Res 25:273–355
Harrap KA, Payne CC, Robertson JS (1977) The properties of three baculoviruses from closely related hosts. Virology 79:14–21
Hess RT, Falcon LA (1978) Electron microscopic observation of the membrane surrounding polyhedral inclusion bodies of insects. Acta Virol 56:169–176
Hink WF, Vail PV (1973) A plaque assay for titration of alfalfa looper nuclear polyhedrosis virus in the cabbage looper (TN-368) cell line. J Invertebr Pathol 22:168–174
Hirumi H, Hirumi K, McIntosh AH (1976) Morphogenesis of a nuclear polyhedrosis virus of the alfalfa looper in a continuous cabbage looper cell line. Annals NY Acad Sci 266:302–326
Howatson AF (1970) Vesicular stomatitis and related viruses. Adv Virus Res 16:195–256
Huber J, Dickler E (1975) Field tests using a granulosis virus to control the codling moth (*Laspeyresia pomonella* [L]). Z Pflanzenkr Pflanzenschutz 82:540–546
Huber J, Dickler E (1977) Codling moth granulosis virus: its efficiency in the field in comparison with organophosphorus insecticides. J Econ Entomol 70:557–561
Huber J, Krieg A (1978) Studies with baculoviruses on mammals. Z Angew Zool 65:69–80
Huger A (1963) Granulosis of insects. In: Steinhaus EA (ed) Insect pathology, vol 1. Academic, New York, pp 531–575
Huger AM (1966) A virus disease of the Indian rhinoceros beetle, *Oryctes rhinoceros* (Linnaeus), caused by a new type of insect virus, Rhabdionvirus oryctes gen, n, sp n. J Invertebr Pathol 8:38–51
Huger A, Kreig A (1961) Electron microscope investigation on the virogenesis of the granulosis of *Choristoneura muriana* (Hubner). J Insect Pathol 3:183–196
Hughes KM (1972) Fine structure and development of two nuclear polyhedrosis viruses. J Invertebr Pathol 19:198–207
Hunter DK (1970) Pathogenicity of a granulosis virus of the Indian meal moth. J Invertebr Pathol 16:339–341
Hunter DK, Hoffman DF (1970) A granulosis virus of the almond moth, *Cadra cautella*. J Invertebr Pathol 16:400–407
Hunter DK, Hoffman DF (1972) Cross infection of a granulosis virus of *Cadra cautella*, with observations on its ultrastructure in infected cells of *Plodia interpunctella*. J Invertebr Pathol 20:4–10
Hunter DK, Hoffman DF (1973) Susceptibility of two strains of Indian meal moth to a granulosis virus. J Invertebr Pathol 20:361–363
Hunter DK, Collier SJ, Hoffman DF (1973) Effectiveness of a granulosis virus of the Indian meal moth as a protectant for stored inshell nuts: preliminary observations. J Invertebr Pathol 22:481
Hunter DK, Hoffman DF, Collier SJ (1975 Observations on a granulosis virus of the potato tuberworm, *Pthorimaea operculella*. J Invertebr Pathol 26:397–400
Hunter DK, Collier SJ, Hoffman DF (1977) Granulosis virus of Indian meal moth as a protectant for stored inshell almonds. J Econ Entomol 70:493–494
Hunter DK, Collier SJ, Hoffman DF (1979) The effect of a granulosis virus on *Plodia interpunctella* (Lepidoptera: Pyralidae) infestations occurring in raisins. J Stored Prod Res 15:65–79
Ignoffo CM (1973) Effects of entomopathogens on vertebrates. Ann NY Acad Sci 217:141–172
Ignoffo CM (1975) Entomopathogens as insecticides. Environ Lett 8:23–40
Ignoffo CM (1979) Viral bioinsecticides. In: Porter JR (ed) Underexploited microbial processes of potential Economic importance. National Academy of Science, Washington DC, pp 142–153
Ito Y, Sakiyama M (1975) Population dynamics of *Pieris rapae crucivora* Boisduval (Lepidoptera: Pieridae), an introduced pest in Okinawa. III. Result of the introduction of *Apanteles glomeratus* and a cabbage butterfly granulosis virus. Jpn J Appl Entomol Zool 19:285–289
Ito Y, Shiga M, Oto N, Nakazawa H (1977). A granulosis virus, possible biological agent for control of *Adoxophyes orana* (Lepidoptera: Tortricidae) in apple orchards. III. A preliminary model for population management. Res Popul Ecol (Kyoto) 19:33–50
Jaques RP, MacLellan CR, Sanford KH, Proverbs MD, Hagley EAC (1977) Preliminary orchard tests on control of codling moth larvae by a granulosis virus. Can Entomol 109:1079–1082
Kawamoto F, Kumada N, Kobayashi M (1977a) Envelopment of the nuclear polyhedrosis virus of the oriental tussock moth *Euproctis subflava*. Virology 77:867–871

Kawamoto F, Suto C, Kumada N, Kobayashi M (1977b) Cytoplasmic budding of a nuclear polyhedrosis virus and comparative ultrastructural studies of envelopes. Microbiol Immunol 21:255–265
Kawanishi CY, Egawa K, Summers MD (1972a) Solubilization of *Trichoplusia ni* granulosis virus proteinic crystal. II. Ultrastructure. J Invertebr Pathol 20:95–100
Kawanishi CY, Summers MD, Stoltz DB, Arnott HJ (1972b) Entry of an insect virus in vivo by fusion of the viral envelope and microvillus membrane. J Invertebr Pathol 20:104–108
Kelly DC (1977) The DNA contained by nuclear polyhedrosis viruses isolated from four *Spodoptera* spp (Lepidoptera: Noctuidae): genome size and homology determined by DNA reassociation kinetics. Virology 76:468–471
Kelly DC (1982) Baculovirus replication. J Gen Virol 63:1–13
Kelly DC, Lescott T (1981) Baculovirus replication: protein synthesis in *Spodoptera frugiperda* cells infected with *Trichoplusia ni* nuclear polyhedrosis virus. Microbiologica 4:35–57
Kelly DC, Lescott T (1983) Baculovirus replication: glycosylation of polypeptides synthesized in *Trichoplusia ni* nuclear polyhedrosis virus infected cells and the effect of tunicamycin. J Gen Virol 64:1915–1926
Kelly DC, Lescott T (1984) Baculovirus replication: phosphorylation of polypeptides synthesized in *Trichoplusia ni* nuclear polyhedrosis virus-infected cells. J Gen Virol 65:1183–1191
Kelly DC, Brown DA, Ayres MD, Allen CJ, Walker IO (1983) Properties of the major nucleocapsid protein of *Heliothis zea* singly enveloped nuclear polyhedrosis virus. J Gen Virol 64:399–408
Kelsey JM (1957) Virus sprays for control of *Pieris rapae* L. NZJ Sci Technol Sect A 38:644–646
Kelsey JM (1958) Control of *Pieris rapae* by granulosis viruses. NZJ Agric Res 1:778–782
Khosaka T, Himeno M, Onodera K (1971) Separation and structure of components of nuclear polyhedrosis virus of the silkworm. J Virol 7:267–273
Kinsinger RA, McGaughey WH (1976) Stabilitiy of *Bacillus thuringiensis* and a granulosis virus of *Plodia interpunctella* on stored wheat. J Econ Entomol 69:149–154
Klenk H-D, Choppin PW (1970) Plasma membrane lipids and parainfluenza virus assembly. Virology 400:939
Knudson DL, Harrap KA (1976) Replication of a nuclear polyhedrosis virus in a continuous cell culture of *Spodoptera frugiperda*: microscopy study of the sequence of events of the virus infection. J Virol 17:254–268
Knudson DL, Tinsley TW (1978) Replication of a nuclear polyhedrosis virus in a continuous cell culture of *Spodoptera frugiperda*: partial characterization of the viral DNA, comparative DNA-DNA hybridization and patterns of DNA synthesis. Virology 87:42–57
Kozlov EA, Alexeenko IP (1967) Electron microscope investigation of the structure of the nuclear polyhedrosis virus of the silkworm, *Bombyx mori*. J Invertbr Pathol 9:413–419
Kozlov EA, Sidorova NM, Serebrayani SB (1975) Proteolytic cleavage of polyhedral protein during dissolution of inclusion bodies of the nuclear polyhedrosis viruses of *Bombyx mori* and *Galleria mellonella* under alkaline conditions. J Invertebr Pathol 25:97–101
Kreig A (1976) Granulosis and nuclear polyhedrosis viruses: safety aspects concerning their production and application. Z Angew Entomol 82:129–134
Kreig A, Huger AM (1969) New ultracytological finding in insect nuclear polyhedrosis. J Invertebr Pathol 13:272–279
Krell RJ, Stoltz DB (1979) Unusual baculovirus of the parasitoid wasp *Apanteles melanoscelus*: isolation and preliminary characterization. J Virol 29:1118–1126
Kurstak E, Bellonick S, Brailovsky C (1969) Transformation de cellules L de souris par un virus d'invertebres: le virus de la densonucleose (VDN). CR Acad Sci 269:1716–1719
Kurstak E, Bellonick S, Garzon S (1970) Immunoperoxydase ultrastructurale: localisation d'antigènes du virus de la densonucleose (VDN) baseé sur les proprietes structurales de la peroxydase. CR Acad Sci 271:2426–2429
Lautenschlager RA, Kircher CH, Podgwaite JD (1977) Effect of nuclear polyhedrosis virus on selected mammalian predators of the gypsy moth. Forest Service Research Paper NE-377. US Department of Agriculture, Washington, DC
Lee HH, Miller LK (1978) Isolation of genotypic variants of *Autographa californica* nuclear polyhedrosis virus. J Virol 27:754–767
Lee HH, Miller LK (1979) Isolation, complimentation and initial characterization of temperature sensitive mutants of the baculovirus *Autographa californica* nuclear polyhedrosis virus. J Virol 31:240–252

Longworth JF, Robertson JS, Payne CC (1972) The purification and properties of inclusion body protein of the granulosis virus of *Pieris brassicae*. J Invertebr Pathol 19:42–50
Lubbert H, Doerfler W (1984) Transcription of overlapping sets of RNA's from the genome of *Autographa californica* nuclear polyhedrosis virus: a novel method for mapping RNA's. J Virol 52:255–265
MacKinnon EA, Henderson JF, Stoltz DB, Faulkner P (1974) Morphogenesis of a nuclear polyhedrosis virus under conditions of prolonged passage in vitro. J Ultrastruct Res 49:419–435
Maeda S, Tanada Y (1983) Protein synthesis of a granulosis virus in the larval fat body of the armyworm, *Pseudaletia unipuncta* (Noctuidae, Lepidoptera). J Invertebr Pathol 41:265–268
Maramarosch K (1968) Insect viruses. Curr Top Microbiol Immunol 42:1–192
Maramarosch K (1977) The atlas of insect and plant viruses. Academic, New York
Martignoni ME (1957) Contributo alla conoscenza di una granulosi di *Eucosma griseana* (Hubner) (Tortricidae: Lepidoptera) quale futtori limitante il pullulamento dellingettonella Engadina alta. Mitt Schweiz Anst Forstt Versuchw 32:371
Martignoni ME, Iwai PJ (1977) A catalog of viral diseases of insects and mites. Forest Service General Technical Report PNW-40. US Department of Agriculture, Washington, DC
Maruniak JE (1979) Biochemical characterization of baculovirus structural and infected TN-368 cell polypeptides, glycoproteins and phosphoproteins. PhD Thesis University of Texas, Austin
Maruniak JE, Summers MD (1978) Comparative peptide mapping of baculovirus polyhedrins. J Invertebr Pathol 32:196–201
Maruniak JE, Summers MD (1981) *Autographa californica* nuclear polyhedrosis virus phosphoproteins and synthesis of intracellular proteins after virus infection. Virology 109:25–34
Maruniak JE, Summers MD, Falcon LA, Smith GE (1979) *Autographa californica* nuclear polyhedrosis virus structural proteins compared from in vivo and in vitro sources. Intervirology 11:82–88
McCarthy WJ, DiCapua RA (1979) Characterization of solubilized proteins from tissue culture and host-derived nuclear polyhedra of *Lymantria dispar* and *Autographa californica*. Intervirology 11:174–181
McCarthy WJ, Liu S (1976) Electrophoretic and serological characterization of *Porthetria dispar* polyhedron protein. J Invertebr Pathol 28:57–65
McGaughey WH (1975a) A granulosis virus for Indian meal moth control in stored wheat and corn. J Econ Entomol 68:346–348
McGaughey WH (1975b) Compatibility of *Bacillus thuringiensis* and granulosis virus treatments of stored grain with four fumigants. J Invertebr Pathol 26:247–250
McGaughey WH, Dickey D, Shesser P (1978) Indian meal moth infestation of farm-stored wheat in Kansas. J Econ Entomol 71:503–506
McIntosch AH, Maramorosch K (1973) Retention of insect virus infectivity in mammalian cell cultures. NY Entomol Soc 81:175–182, 159
McIntosch AH, Shamy R (1975) Effects of the nuclear polyhedrosis virus (NPV) of *Autographa californica* on a vertebrate viper cell line. Ann NY Acad Sci 266:327–331
McIntosch AH, Shamy R (1980) Biological studies of a baculovirus in a mammalian cell line. Intervirology 13:331–341
Merdan A, Croizier L, Veyrunes JC, Croizier G (1977) Etude comparee des proteines des poledres et des virions de trois isolate de baculovirus de *Spodoptera littoralis*. Entomophaga 22:413–420
Miller LK (1981) Construction of a genetic map of baculovirus *Autographa californica* nuclear polyhedrosis virus by marker rescue of temperature sensitive mutants. J Virol 39:973–976
Miller LK, Dawes KP (1978a). Restriction endonuclease analysis for the identification of baculovirus pesticides. Appl Environ Microbiol 35:411–421
Miller LK, Dawes KP (1978b) Restriction endonuclease analysis to distinguish two closely related nucleus polyhedrosis viruses: *Autographa californica* MNPV and *Trichoplusia ni* MNPV. Appl Environm Microbiol 35:1206–1210
Miller LK, Dawes KP (1979) Physical map of the DNA genome of *Autographa californica* nuclear polyhedrosis virus. J Virol 29:1044–1045
Miller LK, Adang MJ, Browne D (1983) Protein kinase activity associated with the extracellular and occluded forms of the baculovirus *Autographa californica* nuclear polyhedrosis virus. J Virol 46:275–278
Miltenburger HG, Naser WL, Harvey JP, Huber J, Huger AM (1984) The cellular substrate: a very important requirement for baculovirus in vitro replication. Z Naturforsch 56:993–1002

Moore NF, Moore JC, Kelly DC (1984) Lipids of nuclear polyhedrosis viruses. Microbiologica 7:267–272

Morgan C, Bergold GH, Moore DH, Rose HM (1955). The macromolecular paracrystalline lattice of insect viral polyhedral bodies demonstrated in ultrathin sections examined in the electron microscope. J Biophys Cytol 1:187–190

Morris TJ, Hess RT, Pinnock DE (1979) Physiochemical characterization of a small RNA virus associated with baculovirus infection in *Trichoplusia ni*. Intervirology 11:238–247

Nagata M, Tanada Y (1983) Origin of an alkaline protease associated with the capsules of a granulosis virus of the armyworm, *Pseudaletia unipuncta*. Arch Virol 76:245–256

Naser WL, Miltenburger HG, Harvey JP, Huber J, Huger AM (1984) In vitro replication of the *Cydia pomonella* (codling moth) granulosis virus. FEMS Microbiol Lett 24:117–121

Paillot A (1926) Sur une nouvelle maladie du noyau au grasserie des chenilles de *P. brassicae* et un nouveau groupe de microorganismes parasites. CR Acad Sci 182:180–182

Paillot A (1934) Un nouveau type de maladie a ultravirus chez les insectes. CR Acad Sci 198:204–205

Payne CC, Compson D, DeLooze SM (1977) Properties of the nucleocapsids of a virus isolated from *Oryctes rhinoceros*. Virology 77:269–280

Pennock GD, Shoemaker C, Miller LK (1984) Strong and regulated expression of *Escherichia coli* β-galactosidase in insect cells with a baculovirus vector. Mol Cell Biol 4:399–406

Pinnock DE, Hess RT (1978) Morphological variation in the cytopathology associated with granulosis virus in the fruit-tree leafroller, *Archips argyraspila*. J Ultrastruct Res 63:252–260

Potter KN, Faulkner P, MacKinnon EA (1976) Strain selection during serial passage of *Trichoplusia ni* nuclear polyhedrosis virus. J Virol 18:1040–1050

Potter KN, Jaques RP, Faulkner P (1978) Modification of *Trichoplusia ni* nuclear polyhedrosis virus passaged in vivo. Intervirology 9:76–85

Pritchett DW, Young SY, Yearian WC (1984) Some factors involved in the dissolution of *Autographa californica* nuclear polyhedrosis virus polyhedra by digestive fluids of *Trichoplusia ni* larvae. J Invertebr Pathol 43:160–168

Printz P (1973) Relationship of sigma virus to vesicular stomatitis virus. Adv Virus Res 18:143–157

Raghow R, Grace TDC (1974) Studies on a nuclear polyhedrosis virus in *Bombyx mori* cells *in vitro*. I. Multiplication kinetics and ultrastructural studies. J Ultrastruct Res 47:384–399

Reinganum C, Robertson JS, Tinsley TW (1978) A new group of RNA viruses from insects. J Gen Virol 40:195–202

Rivers CF (1976) Some problems in the use of insect viruses as pesticidal agents in the control of pest populations. Z Angew Entomol 82:148–152

Roberts DW, Granados RR, McCarthy WJ (1973) Interactions of entomopoxviruses and vaccinia virus. Abstracts, International Colloquium on Insect Pathology and Microbiological Control, Oxford, England p 29

Robertson JS, Harrap KA, Longworth JF (1974) Baculovirus morphogenesis: the acquisition of the virus envelope. J Invertebr Pathol 23:248–251

Röder A, Pünter J (1977) Interactions between nuclear polyhedrosis viruses and vertebrate cells. Zentralbl Bakteriol Parasitenkd Infektionskr Hyg Abt 1 Orig Reihe A 239:459–464

Rohrmann GF, McParland RH, Martignoni ME, Beandreau GS (1978) Genetic relatedness of two nucleopolyhedrosis viruses pathogenic for *Orgyia pseudotsugata*. Virology 84:213–217

Rubinstein R, Lawler KA, Granados RR (1982) Use of primary fat body cultures for the study of baculovirus replication. J Invertebr Pathol 40:266–273

Russell DL (1986) An investigation of the structural proteins of the granulosis virus infecting *Plodia interpunctella*. PhD thesis. Kansas State University, Manhattan

Russell, DL, Consigli RA (1985) Glycosylation of purified enveloped nucleocapsids of the granulosis virus infecting *Plodia interpunctella* as determined by lectin binding. Virus Res 4:83–91

Sato T, Otto N, Nurita N (1977) A granulosis virus of the peach fruit moth, *Carposina niponensis* Walsingham (Lepidoptera: Carposinidae). Appl Entomol Zool 12:365–369

Shapiro M, Ignoffo CM (1970) Hemagglutination by a nucleopolyhedrosis virus of the cotton bollworm, *Heliothis zea*. Virology 41:577–579

Shieh TR, Bohmfalk GT (1980) Production and efficacy of baculoviruses. Biotechnol Bioeng 22:1357–1375

Shvedchikova NG, Tarasevich LM (1968) On the biochemistry of virus granulosis of *Dendrolimus sibiiricus* (Tschetw). Abstracts 13th International Congress for Entomology, Moscow, p 244

Shvedchikova NG, Tarasevich LM (1971) Electron microscope investigation of granulosis viruses of *Dendrolimus sibiricus* and *Agrotis segetum*. J Invertebr Pathol 18:25–32
Singh KRP (1972) Growth of arboviruses in arthropod tissue culture. Adv Virus Res 17:187–206
Singh SP, Gudauskas RT, Harper JD (1983) High resolution two-dimensional gel electrophoresis of structural proteins of baculoviruses of *Autographa californica* and *Porthetria* (*Lymantria*) *dispar*. Virology 125:370–380
Singh SP, Gudauskas RT, Harper JD, Edwards J (1985) Two-dimensional gel electrophoresis of basic proteins of *Autographa californica* nuclear polyhedrosis virus. J Invertebr Pathol 45:249–251
Smith GE, Summers MD (1978) Analysis of baculovirus genomes with restriction endonucleases. Virology 89:517–527
Smith GE, Summers MD (1979) Restriction maps of five *Autographa californica* MNPV variants, *Trichoplusia ni* MNPV, and *Galleria mellonella* MNPV DNAs with endonucleases *Sma*I, *Kpn*I, *Bam*H1, *Sac*I, *Xho*I, and *Eco*RI. J Virol 30:828–838
Smith GE, Summers MD (1980) Restriction map of *Rachiplusia ou* and *Rachiplusia ou – Autographa californica* baculovirus recombinants. J Virol 33:311–319
Smith GE, Summers MD, Fraser MJ (1983) Production of human beta interferon in insect cells infected with a baculovirus expression vector. Mol Cell Biol 3:2156–2165
Smith KM (1976) Virus-insect relationships. Longman, New York
Smith KM, Hills GJ (1962) Multiplication and ultrastructure of insect viruses. In: Proceedings of the 11th International Congress for Entomology, vol 2. Vienna, Austria, pp 823–827
Stairs GR, Parrish WB, Briggs JD, Allietta M (1966) Fine structure of a granulosis virus of the codling moth. Virology 30:583–584
Steinhaus EA (1947) A new disease of the variegated cutworm, *Peridroma margaritosa* (Haw). Science 106:323
Stiles B, Wood HA (1983) A study of the glycoproteins of *Autographa californica* nuclear polyhedrosis virus (AcNPV). Virology 131:230–241
Stiles B, Wood HA, Hughes PR (1983) Effect of tunicamycin on the infectivity of *Autographa californica* nuclear polyhedrosis virus. J Invertebr Pathol 41:405–408
Stoltz DB, Vinson SB (1979) Viruses and parasitism in insects. Adv Virus Res 24:125–171
Stoltz DB, Pavan C, daCunha A (1973) Nuclear polyhedrosis virus: a possible example of *de novo* intranuclear membrane morphogenesis. J Gen Virol 19:145–150
Summers MD (1969) Apparent in vivo pathway of granulosis virus invasion and infection. J Virol 4:188–190
Summers MD (1971) Electron microscopic observations on granulosis virus entry, uncoating and replication processes during infection of the midgut cells of *Trichoplusia ni*. J Ultrastruc Res 35:606–625
Summers MD (1975) Biophysical and biochemical properties of baculoviruses. In: Summers M, Engler R, Falcon LA, Vail P (eds) Baculoviruses for insect pest control: safety considerations. American Society for Microbiology, Washington DC, pp 17–29
Summers MD (1977) Baculoviruses (Baculoviridae). In: Maramorosch K (ed) The atlas of insect and plant viruses. Academic, New York, pp 3–27
Summers MD, Anderson DL (1972a) Granulosis virus deoxyribonucleic acid: a closed double-stranded molecule. J Virol 9:710–713
Summers MD, Anderson DL (1972b) Characterization of deoxyribonucleic acid isolated from the granulosis viruses of the cabbage looper, *Trichoplusia ni* and the fall armyworm, *Spodoptera frugiperda*. Virology 50:459–471
Summers MD, Egawa K (1973) Physical and chemical properties of *Trichopulsia ni* granulosis virus granulin. Virology 12:1092–1103
Summers MD, Paschke JD (1970) Alkali-liberated granulosis virus of *Trichoplusia ni*. I. Density gradient purification of virus components and some of their in vitro chemical and physical properties. J Invertebr Pathol 16:227–240
Summers MD, Smith GE (1975a) Comparative studies of baculovirus granulins and polyhedrins. Intervirology 6:168–180
Summers MD, Smith GE (1975b) *Trichoplusia ni granulosis* virus granulin: a phenol-soluble, phosphorylated protein. J Virol 16:1108–1116
Summers MD, Smith GE (1978) Baculovirus structural polypeptides. Virology 84:390–402
Summers MD, Volkman LE (1976) Comparison of biophysical and morphological properties of

occluded and extracellular non-occluded baculovirus from in vivo and in vitro host systems. J Virol 17:962–972

Summers MD, Smith GE, Knell JD, Burand JP (1980) Physical maps of *Autographa californica* and *Rachiplusia ou* nuclear polyhedrosis virus recombinants. J Virol 34:693–703

Tanada Y, Leutenegger R (1970) Multiplication of a granulosis virus in larval midgut cells of *Trichoplusia ni* and possible pathways of invasion into the hemocoel. J Ultrastruct Res 330:589–600

Tanada Y, Watanabe H (1971) Disc electrophoresis and serological studies of the capsule proteins obtained from two strains of a granulosis virus of the armyworm, *Pseudaletia unipuncta*. J Invertebr Pathol 18:307–312

Tanada Y, Hess RT, Omi EM (1984) The movement and invasion of an insect baculovirus in tracheal cells of the armyworm. *Pseudaletia unipuncta*. J Invertebr Pathol 44:198–208

Teakle RE (1969) A nuclear polyhedrosis virus of *Anthela varia* (Lepidoptera: Anthelidae). J Invertebr Pathol 14:18–27

Tjia ST, Carstens EB, Doerfler W (1979) Infection of *Spodoptera frugiperda* cells with *Autographa californica* nuclear polyhedrosis virus. II. The viral DNA kinetics of its replication. Virology 99:399–409

Tweeten KA, Bulla LA Jr, Consigli RA (1977a) Isolation and purification of a granulosis virus from infected larvae of the Indian meal moth, *Plodia interpunctella*. Appl. Environ Microbiol 34:320–327

Tweeten KA, Bulla LA Jr, Consigli RA (1977b). Supercoiled circular DNA of an insect granulosis virus. Proc Natl Acad Sci USA 74:3574–3578

Tweeten KA, Bulla LA Jr, Consigli RA (1978) Characterization of an alkaline protease associated with a granulosis virus of *Plodia interpunctella*. J Virol 26:702–711

Tweeten KA, Bulla LA Jr, Consigli RA (1980a) Characterization of an extremely basic protein derived from granulosis virus nucleocapsids. J Virol 33:866–876

Tweeten KA, Bulla LA Jr, Consigli RA (1980b). Restriction enzyme analysis of the genomes of *Plodia interpunctella* and *Pieris rapae* granulosis viruses. Virology 104:514–519

Tweeten KA, Bulla LA Jr, Consigli RA (1980c). Structural polypeptides of the granulosis virus of *Plodia interpunctella*. J Virol 33:877–886

Tweeten KA, Bulla LA Jr, Consigli RA (1981) Applied and molecular aspects of insect granulosis viruses. Microbiological Rev. 45:379–408

Vago C, Bergoin M (1963) Development des virus a corps d'inclusion du lepidoptere *Lymantria dispar* en cultures cellulaires. Entomophaga 8:253

Vlak JM, Smith GE (1982) Orientation of the genome of *Autographa californica* nuclear polyhedrosis virus: a proposal. J Virol 41:1118–1121

Volkman LE, Summers MD, Hsieh CH (1976) Occluded and non-occluded nuclear polyhedrosis virus grown in *Trichoplusia ni*: comparative neutralization, infectivity and in vitro growth studies. J Virol 19:820–832

Wager R, Benz G (1971) Histochemical studies on nucleic acid metabolism in granulosis-infected *Carpocapsa pomonella*. J Invertebr Pathol 17:107–115

Walker S, Kawanishi CY, Hamm JJ (1982) Cellular pathology of a granulosis virus infection. J Ultrastruct Res 80:163–177

Watanabe H, Kobayashi M (1970) Histopathology of a granulosis in the larva of the fall webworm, *Hyphantria cunea*. J Invertebr Pathol 16:71–79

Whitfield SG, Murphy FA, Sudia WD (1973) St. Louis encephalitis virus: an ultrastructural study of infection in a mosquito vector. Virology 56:70–87

Wildy P (1971) Classification and nomenclature of viruses. Monogr Virol 5:1–81

Wilson F (1960) The effectiveness of a granulosis virus applied to field populations of *Pieris rapae* (Lepidoptera). Aust J Agric Res 2:485–497

Wilson ME, Consigli RA (1984) Polyamine content of the *Plodia interpunctella* granulosis virus and its larval host. J Invertebr Pathol 44:161–165

Wilson ME, Consigli RA (1985a) Characterization of a protein kinase activity associated with purified capsids of the granulosis virus infecting *Plodia interpunctella*. Virology 143:516–525

Wilson ME, Consigli RA (1985b) Functions of a protein kinase activity associated with purified capsids of the granulosis virus infecting *Plodia interpunctella*. Virology 143:526–535

Wood HA (1977) An agar overlay plaque assay method for *Autographa californica* nuclear polyhedrosis virus. J Invertbr Pathol 29:304:307

Wood HA (1980a) *Autographa californica* nuclear polyhedrosis virus induced proteins in tissue culture. Virology 102:21–27
Wood HA (1980b) Isolation and replication of an occlusion body-deficient mutant of the *Autographa californica* nuclear polyhedrosis virus. Virology 105:338–344
Wyatt GR (1952a) Specificity in the composition of nucleic acids. Exp Cell Res Suppl 2:201–217
Wyatt GR (1952b) The nucleic acids of some insect viruses. J Gen Physiol 36:201–205
Xeros N (1956) The virogenic stroma in nuclear and cytoplasmic polyhedroses. Nature (London) 128:412–413
Yamafugi K, Makai J, Yoshihara F (1959) Deoxyribonuclease and protease in polyhedral viral particles. Enzymologia 22:1–12
Yamamoto T, Tanada Y (1978a) Biochemical properties of viral envelopes of insect baculoviruses and their role in infectivity. J Invertebr Pathol 32:202–211
Yamamoto T, Tanada Y (1978b) Protein components of two strains of granulosis virus of the armyworm *Pseudaletia unipuncta* (Lepidoptera: Noctuidae). J Invertebr Pathol 32:158–170
Yamamoto T, Tanada Y (1979) Comparative analysis of the enveloped virions of two granulosis viruses of the armyworm, *Pseudaletia unipuncta*. Virology 94:71–81
Zettler JL, McDonald LL, Redlinger LM, Jones RD (1973) *Plodia interpunctella* and *Cadra cautella* resistance in strains to malathion and synergized pyrethins. J Econ Entmol 66:1049–1050
Zummer M, Faulkner P (1979) Absence of protease in baculovirus polyhedral bodies propagated in vitro. J Invertebr Pathol 33:383–384

The 64K Envelope Protein of Budded *Autographa californica* Nuclear Polyhedrosis Virus

L.E. VOLKMAN

1 Introduction

At least 19 published studies describe the number and molecular weights of baculovirus structural proteins (reviewed by PADHI 1985; VOLKMAN 1983; SINGH et al. 1985). By far the majority of these papers (ten) feature the structural proteins of *Autographa californica* nuclear polyhedrosis virus (AcNPV). Some of the studies additionally examine the temporal expression of baculovirus-induced infection-specific proteins. Most agree that baculoviruses are composed of about 20–30 proteins, that roughly the same number of proteins are specifically induced during infection, and that groups of infection-specific proteins are produced at different times postinfection (CARSTENS et al. 1979; DOBOS and COCHRAN 1980; WOOD 1980; KELLY and LESCOTT 1981; and see CARSTENS and FAULKNER, this volume). A few of these studies include experiments aimed at determining which of the proteins are glycosylated (DOBOS and COCHRAN 1980; GOLDSTEIN and MCINTOSH 1980; STILES and WOOD 1983; KELLY and LESCOTT 1983) and which are phosphorylated (DOBOS and COCHRAN 1980; MARUNIAK and SUMMERS 1981). Detailed information regarding most baculovirus proteins beyond this basic level, however, is lacking. There are three exceptions: the inclusion body proteins polyhedrin and granulin (for excellent reviews see VLAK and ROHRMANN 1985, and KOZLOV, this volume), the major nucleocapsid DNA-binding protein (TWEETEN et al. 1980; KELLY et al. 1983), and the 64K

Department of Entomology and Parasitology, University of California, Berkeley, CA 94720, USA

envelope protein of budded AcNPV. While most is known about the inclusion body proteins, the latter two have been examined in enough detail to gain an understanding of their basic chemical properties, their location within the virion, and their probable function. This chapter traces the development of our knowledge about the 64K envelope protein of budded AcNPV.

2 Two Phenotypes

It has been recognized for a number of years that baculoviruses of subgroups A and B are (so far) unique among viruses in that two phenotypes are involved in their persistence in nature: the occluded form that matures within the nucleus where it becomes enveloped and embedded in a protein matrix of polyhedrin or granulin, and the extracellular form that matures by budding through the plasma membrane. The occluded form of the virus, protected by the protein matrix, persists in the soil or on plant foliage until it is ingested by a susceptible larva. In the midgut the alkaline digestive juices dissolve the occluding matrix and liberate the enveloped virions, which are then free to infect their specific target, the columnar epithelial cells. The virus replicates in the nucleus of these cells and progeny virions are produced which bud from the basement membrane into the hemocoel. These budded virions go on to infect other tissues which produce both forms of the virus (Fig. 1).

The pathology of the infection is typically quite severe. The host dies and often liquifies, dispersing polyhedra or capsules into the environment. Recent reviews describing these events in more detail are available (Granados 1980; Faulkner 1981; Tanada and Hess 1984; Vaughn and Dougherty 1985).

Replication of a few of the NPVs has been studied in cell culture. Development of the two forms of virus *in vitro* is clearly temporally regulated, with the budded form being produced first (Volkman et al. 1976). Cells of commonly used lepidopteran cell lines TN-368 and IPLB-21 show characteristic (CPE) upon infection with AcNPV. The progressive change in the character of the CPE reflects stages in virus replication (Volkman and Summers 1975; Volkman et al. 1976). Rounded cells with condensed areas of virogenic stroma in the nucleus appear with the onset of virus budding at about 10 h pi (Fig. 2A, B). In cultures infected at an MOI of 5 or more, the number of cells showing this type of CPE rapidly increase until about 16 h pi, at which time they begin to be replaced by cells containing occluded virus, or polyhedra (Fig. 2D, E) (Volkman et al. 1976).

In addition to differing in both location and timing of maturation, the two forms of the virus differ morphologically – especially the M (multiple nucleocapsids per envelope) NPVs. The polyhedra-derived virus (PDV) of the MNPVs is composed of varying numbers of nucleocapsids wrapped in tightly adhering envelopes that appear to be devoid of surface projections or peplomers (Fig. 2F). In contrast, the budded virus (BV) is composed primarily of single nucleocapsids per envelope, and the envelopes are relatively loose-fitting and contain peplomers (Summers and Volkman 1976; Adams et al. 1977; and Fig. 2B, C).

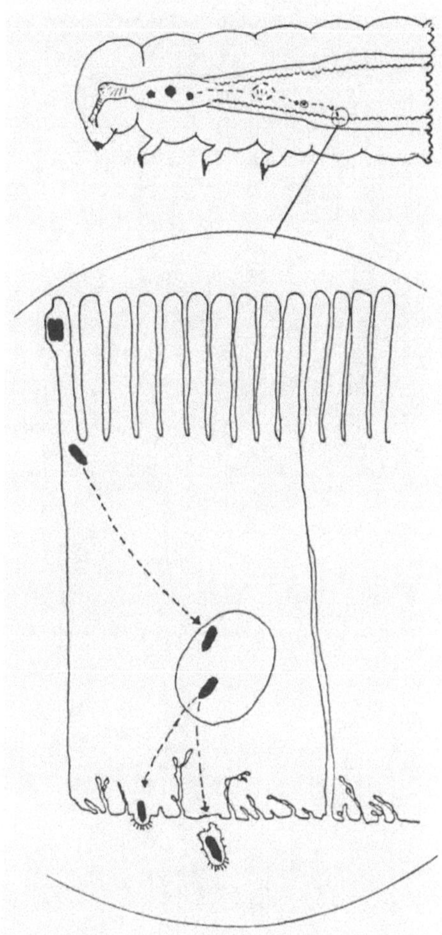

Fig. 1. The natural infection cycle of subgroup A and B baculoviruses begins when viral inclusion bodies are ingested by susceptible larvae. The occluded virus is released in the midgut by dissolution of the inclusion bodies by alkaline gut juices. The released virus infects the columnar epithelial cells of the midgut. Viral replication takes place in the nucleus of these cells, and progeny virus (of a different phenotype than the originally infecting virus) buds from the basement membrane into the hemocoel. This budded virus causes a systemic infection which results in the production of both the occluded and budded viral phenotypes

Studies aimed toward detecting biological differences in the two forms of AcNPV have demonstrated that the budded form of the virus is about a factor of 1800 more infectious in vitro in TN-368 cells than the polyhedra derived form (VOLKMAN et al. 1976). Their infectivities in vivo differ also, and depend on the portal of entry (VOLKMAN and SUMMERS 1977; KEDDIE and VOLKMAN 1985). In 4th instar *Trichoplusia ni* larvae, AcNPV BV is two orders of magnitude more infectious than PDV if injected into the hemocoel. If injected into the gut, PDV is the more highly infectious by three to four orders of magnitude.

3 The 64K Protein of AcNPV

In view of the fact that the two forms of the virus gain their envelopes in different locations (nucleus and plasma membrane), vary widely in infectivity,

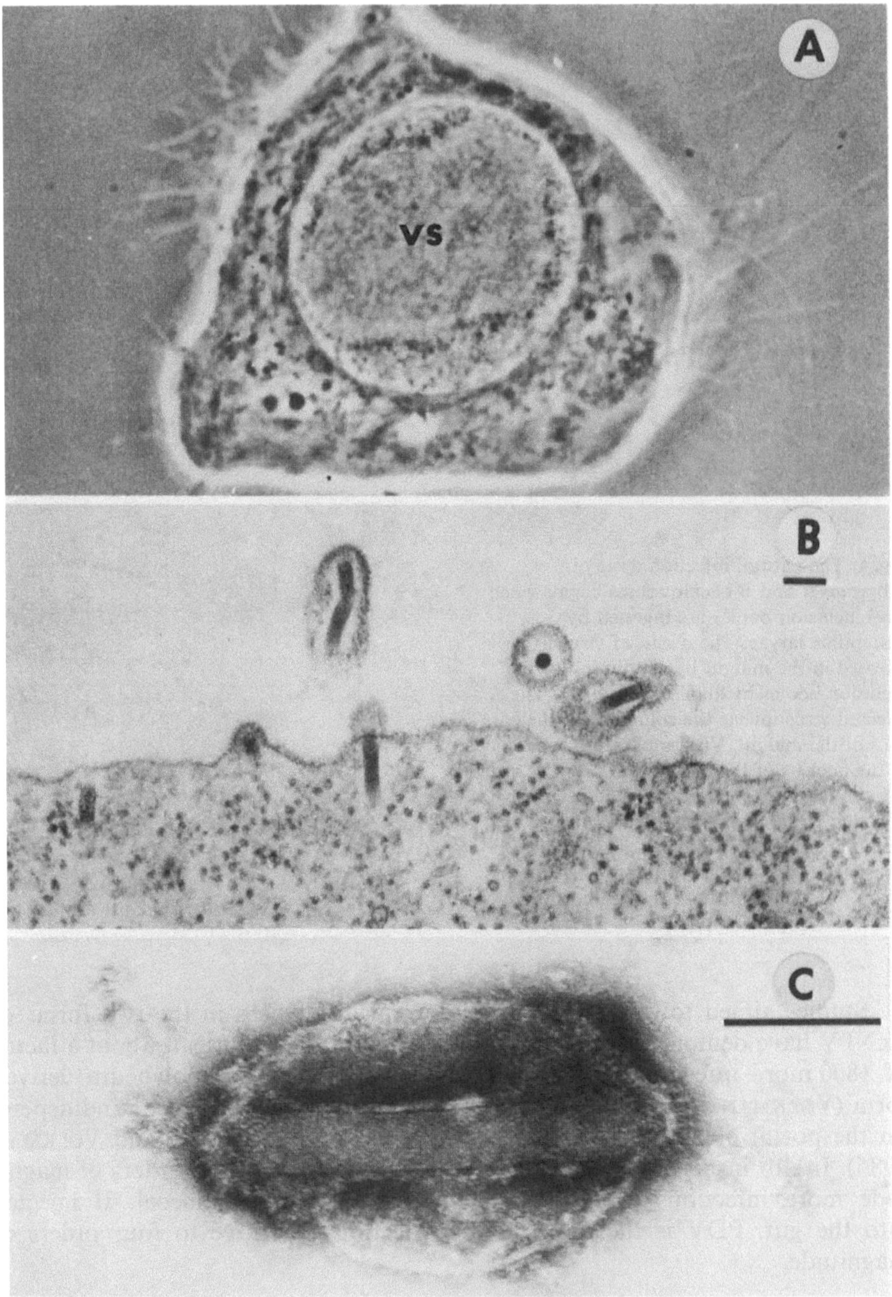

Fig. 2. A The onset of budding of AcNPV in an infected TN-368 cell occurs coincident with the rounding of the normally spindle-shaped cell and the appearance of virogenic stroma (*vs*) in the nucleus. **B** AcNPV budding from the plasma membrane of such an infected cell. **C** AcNPV-budded virion with terminal peplomers. **D** Later in infection the nucleus contains numerous polyhedra. **E** These polyhedra contain bundles of enveloped virions **F** which can be released by exposure to dilute alkaline saline. The released virions are packaged in tightly adhering envelopes that appear to be devoid of surface projections. Empty capsids are also seen (*arrows*). The *bars* in **B**, **C**, **E**, and **F** represent 100 nm

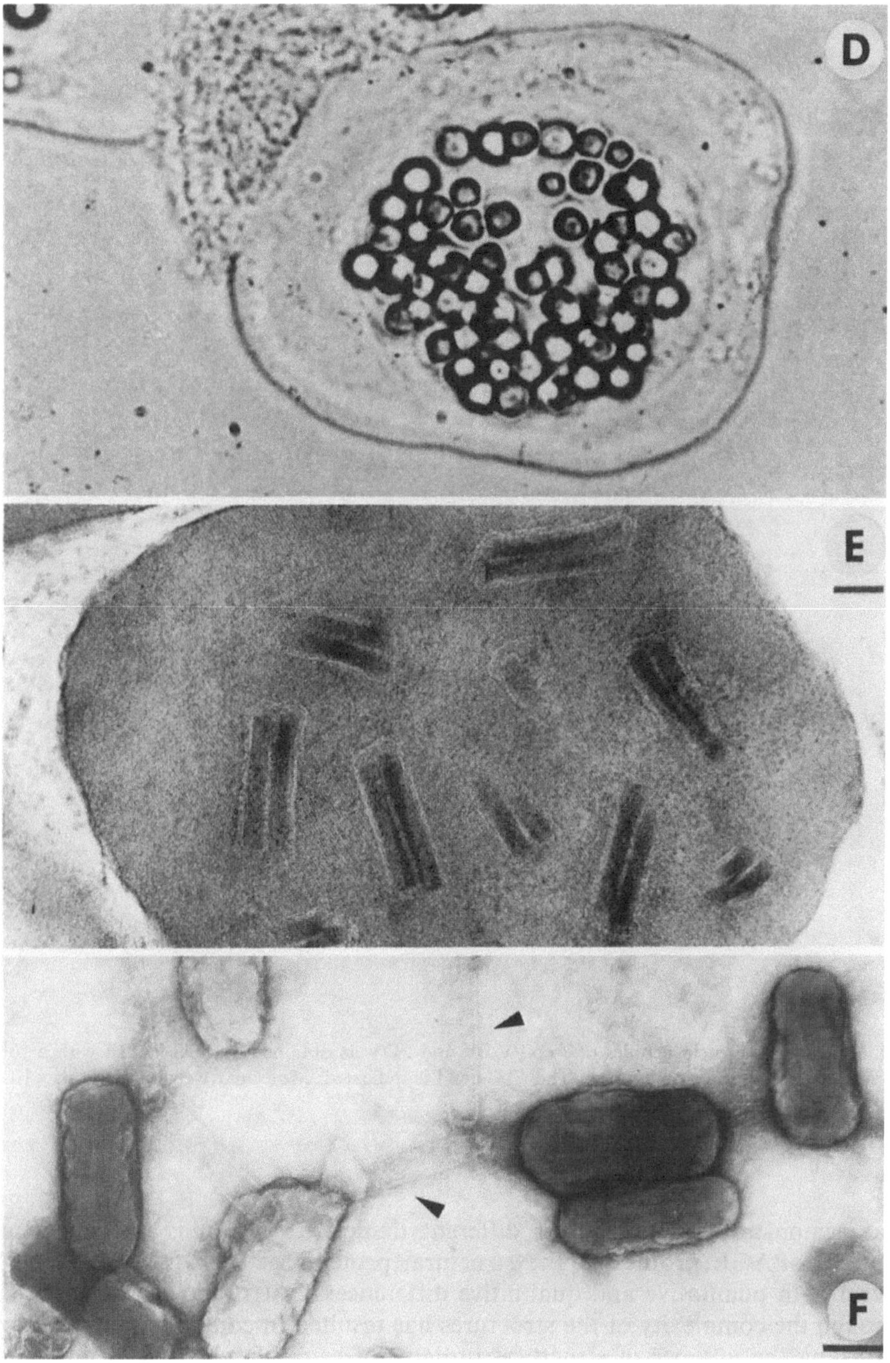
D
E
F

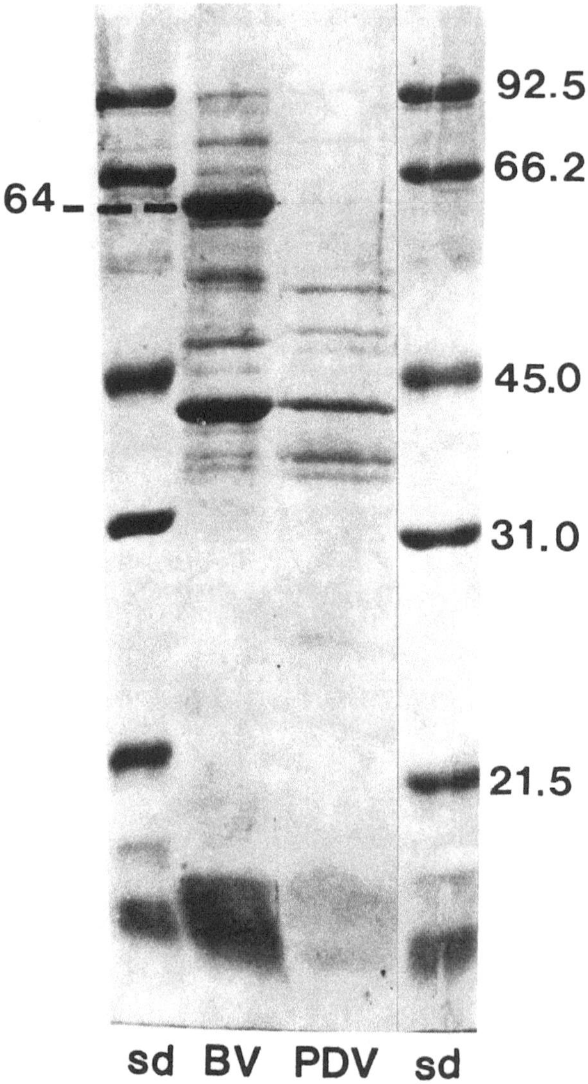

Fig. 3. Structural protein profiles of AcNPV BV and PDV as obtained by SDS-PAGE with a 10% gel and stained with Comassie blue. The 64K band is indicated. Molecular weight standards $\times 10^{-3}$ (*sd*)

and demonstrate specificity for different tissues in vivo, it is not surprising that SDS-PAGE profiles of the structural proteins of each form of AcNPV reveal both quantative and qualitative differences (SMITH and SUMMERS 1978). Though the complexity of the structures has resulted in conflicting information regarding the number of structural proteins of each phenotype (PADHI 1985), it is agreed that one major difference between the budded and occluded forms is that the budded form has considerably more 64K[1] protein on a per weight basis (Fig. 3). Whether the 64K proteins of the two forms are the same molecular species is still an unresolved question and will be addressed in a later section.

3.1 Neutralization, In Vitro and In Vivo

As might be expected, the structural and infectivity differences in BV and PDV are reflected by antigenic differences. Reciprocal neutralization experiments performed in vitro show that antiserum raised to PDV neutralizes the homologous virus, but not the heterologous budded form of the virus; antiserum raised to BV neutralizes both forms (VOLKMAN et al. 1976; ROBERTS 1983). These results indicate that BV has a surface antigen important in its infectivity that is not a component of PDV.

Attempts to determine which structural protein(s) was involved in eliciting the neutralizing antibody specific for BV were made by raising monospecific antibodies to bands of BV structural proteins cut out of SDS-polyacrylamide gels, then assaying the antisera for neutralizing activity (M.D. SUMMERS, personal communication). The 64K BV protein was one promising candidate, but unfortunately these attempts were unsuccessful due to the lack of demonstrable neutralizing activity.

In 1983 HOHMANN and FAULKNER reported that they had generated monoclonal antibodies to budded AcNPV, and that 4 out of 28 showed some degree of neutralizing activity for BV. They also reported that none of the four antibodies reacted with PDV as assessed by ELISA, although titers of 2000 to 16000 were recorded when BV was used as the antigen. Their attempts to identify the BV-reactive polypeptide(s) by Western blot analysis failed, however. The target antigen of one of these neutralizing monoclonal antibodies, AcV_1, was eventually determined by radioimmune precipitation and was found to be the major 64K BV protein (VOLKMAN et al. 1984).

In vitro neutralization experiments have demonstrated that the infectivity of BV preparations can be reduced by three to four orders of magnitude, but not completely eliminated, by exposure to affinity-purified AcV_1. A pleateau is reached where addition of more AcV_1 does not reduce the infectivity further (VOLKMAN and GOLDSMITH 1985). This condition of antibody excess occurs at 2 μg or more of AcV_1 per 1×10^6 PFU BV. Experiments aimed at discovering the nature of these neutralization-resistant virions revealed that they do not lack the AcV_1-reactive epitope, but rather are infectious even though they are linked to antibody (VOLKMAN et al., in press).

In vivo neutralization experiments with AcV_1 and AcNPV BV have been conducted by KEDDIE and VOLKMAN (1985), who demonstrated that the infectivity of budded virus can be reduced by two orders of magnitude if it is incubated with AcV_1 prior to injection into the hemocoel of 4th instar *Trichoplusia ni* larvae. If AcV_1-treated BV is injected into the gut, no neutralization is appar-

[1] The true MW of this protein probably will not be known until it is sequenced. Estimates of it's MW by SDS-PAGE usually vary between 64K DA and 70K DA, and thus care must be taken when comparing studies to ensure that the data refer to the *same* protein.

ent. These results suggest that the 64K protein plays an important role in AcNPV BVs infection of natural host tissue, and provides some assurance that information obtained about the role of 64K in vitro reflects the situation in vivo.

A hint of the nature of the AcV_1-reactive epitope of the 64K protein was obtained from two observations. First, the successful AcV_1-immunoprecipitation of a 59K protein from solubilized, tunicamycin grown, [^{35}S] methionine-labeled BV indicates that the epitope is composed of protein and not carbohydrate (VOLKMAN et al. 1984). Second, the loss of immunoprecipitability of 64K by AcV_1 upon its exposure to 1% SDS demonstrates that the AcV_1-reactive epitope is SDS labile (VOLKMAN and GOLDSMITH 1984). The latter observation explains why monospecific antiserum made to SDS-PAGE purified 64K protein does not neutralize BV, as well as why the target of AcV_1 cannot be detected by Western blotting.

3.2 Time of Synthesis, Phosphorylation, and Glycosylation

CARSTENS et al. (1979) were the first to conduct a study aimed at determining the temporal expression of AcNPV-induced proteins in infected cells. They judged the major BV protein to be 67K, and determined that in a *Spodoptera frugiperda* cell line infected at an MOI of 10, a 67K protein could first be detected at 6–8 h pi. Synthesis of the 67K protein peaked at 12–16 h pi, and very little, if any, could be detected after 24 h pi. The authors noted that even though the 67K protein was a major BV component, it was a relatively minor species in infected cell extracts. In a similar study of AcNPV-infected *Trichoplusia ni* cells, WOOD (1980) first detected the appearance of a 67K protein at 12 h pi, a time coincident with nuclear swelling and the first maturation of BV. Wood detected the onset of production of other infection-specific proteins both before and after the initial appearance of the 67K protein. DOBOS and COCHRAN (1980), studying AcNPV-infected *Spodoptera frugiperda* cells, noted a prominent 66K protein in BV whose synthesis they first detected at 12–18 h pi – in a middle group with regard to time of initial appearance. MARUNIAK and SUMMERS (1981), studying phosphoprotein synthesis in AcNPV-infected TN-368 cells, found that the major 64K BV protein was phosphorylated and could be detected in the cytoplasm by 15 h pi. GOLDSTEIN and MCINTOSH (1980) reported a 70K glycosylated protein as a major AcNPV BV component, which was later confirmed by STILES and WOOD (1983) as a glycoprotein of 67K (1983). Taken together, these studies are in agreement that a 64–70K protein is a major structural component of AcNPV BV, and is synthesized during an intermediate time during infection. Also, evidence was obtained that the major BV protein is both phosphorylated and glycosylated.

3.3 Biochemical Nature

Radioimmune precipitation with AcV_1 of solubilized BV antigens labeled with *N*-acetyl-D-[1^{3}H] glucosamine (VOLKMAN et al. 1984) and $^{32}PO_4$ (VOLKMAN and

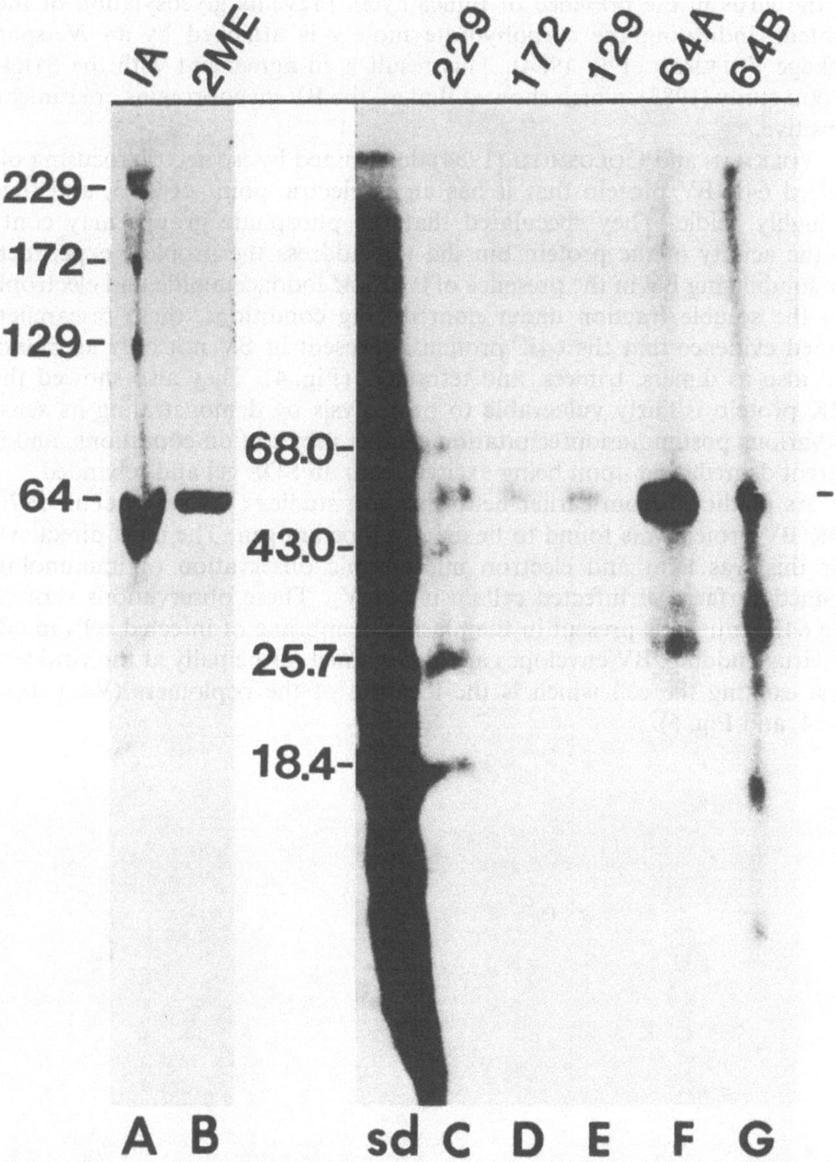

Fig. 4. Multimers of the 64K protein are detected when an AcV_1-immunoprecipitated sample of $[S^{35}]$methionine-labeled solubilized BV is boiled for 3 min in a sample buffer containing 0.075 *M* iodoacetamide (*A*) instead of the usual 3% 2-mercaptoethanol (*B*). The multimers, 229K (*C*), 172K (*D*), and 129K (*E*) are reduced to 64K upon excision from the gel shown in *A*, boiling in sample buffer containing 2-mercaptoethanol, and rebanding by SDS-PAGE. The 64K monomer becomes somewhat degraded upon excision and rebanding; *F*, the 64K band from *A*; *G*, the 64K band from *B*. Molecular weight standards $\times 10^{-3}$ (*sd*)

GOLDSMITH 1984) confirmed the 64K protein is a phosphoglycoprotein. Growth of the virus in the presence of tunicamycin prevents glycosylation of the 64K protein, indicating the carbohydrate moiety is attached by an *N*-asparagine linkage VOLKMAN et al. 1984). This result is in agreement with the STILES and WOOD study (1983), which showed that all the BV glycoproteins are tunicamycin sensitive.

VOLKMAN and GOLDSMITH (1984) determined by isoelectric focusing of solubilized 64K BV protein that it has an isoelectric point of 3.15, and therefore is highly acidic. They speculated that the phosphate groups may contribute to the acidity of the protein but did not address the problem experimentally. By solubilizing BV in the presence of 0.075 *M* iodoacetamide and electrophoresing the soluble fraction under nonreducing conditions, these researchers obtained evidence that the 64K protein is present in BV not only as monomers, but also as dimers, trimers, and tetramers (Fig. 4). They also showed that the 64K protein is fairly vulnerable to proteolysis by demonstrating its sensitivity to various postimmunoprecipitation sample preparation conditions, and its apparent degredation upon being excised from an SDS gel and rebanded.

As predicted from earlier neutralization studies (VOLKMAN et al. 1976), the 64K BV protein was found to be an envelope antigen. The most direct evidence for this was light and electron microscopic observation of immunologically stained surfaces of infected cells using AcV_1. These observations showed that the 64K antigen is present in the plasma membrane of infected cells in advance of virus budding. BV envelopes are also stained, principally at the viral terminus first exciting the cell which is the location of the peplomers (VOLKMAN et al. 1984, and Fig. 5).

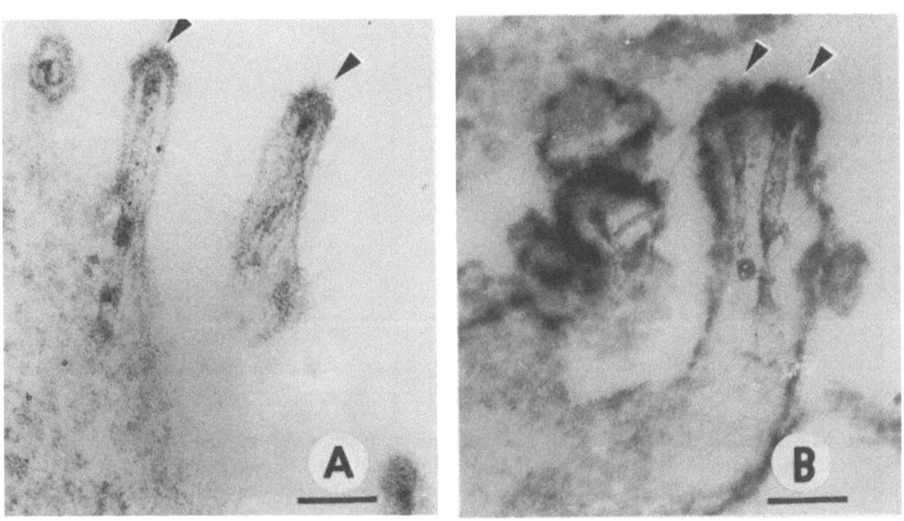

Fig. 5A, B. Immunoelectron micrographs of portions of AcNPV-infected IPLB-21 cells 28 h pi subjected to peroxidase antiperoxidase staining. **A** Cell treated with undiluted DMEM culture medium alone; **B** cell treated with a 1:50 dilution of AcV_1-containing DMEM. *Arrows* indicate peplomers on budding virions in **A**, and an intense region of staining in **B**. The *bar* represents 100 nm

Most of the characteristics of the 64K protein are common among peplomer or spike proteins of enveloped viruses. Peplomer proteins are major envelope glycoproteins, frequently composed of dimers or trimers, and often the target of neutralizing antibodies (DOBOIS-DALCQ et al. 1984). Although not conclusively demonstrated, the 64K protein is very likely the substance of budded virus peplomers.

3.4 Role in Infectivity

The physical and chemical similarity of the 64K BV protein to peplomer proteins of other viruses suggests that it might play a similar role in infection, i.e., in adsorption and/or penetration. To date studies aimed at determining the role of the 64K protein in BV infectivity consist of the examination of various aspects of neutralization, and the mechanism of functional viral entry (VOLKMAN and GOLDSMITH 1985).

The question of whether the 64K protein is involved in virus adsorption has not as yet been fully examined. Whether AcV_1 neutralizes BV by preventing adsorption has been addressed by two different approaches: by comparing AcV_1 and normal immunoglobulin-treated [^{35}S]methionine-labeled BV binding to host cells, and by assessing the neutralizing capability of AcV_1 on virus preadsorbed to host cells. Results of these two different experimental approaches show that AcV_1 does not inhibit binding of BV to cells, and that BV preadsorbed to host cells is still neutralizable by AcV_1 (although not to the extent that nonadsorbed BV is) (VOLKMAN and GOLDSMITH 1985). These results indicate that adsorption is not the step being inhibited by AcV_1, although they do not preclude a role for the 64K protein in adsorption.

Baculovirus entry studies have typically been conducted by electron microscopy (RAGHOW and GRACE 1974; HIRUMI et al. 1975; ADAMS et al. 1977; BASSEMIR et al. 1983) and, recently, by biochemical analysis (WANG and KELLY 1985). These methods have the disadvantage of not being able to distinguish infectious from noninfectious particles. This limitation is serious with BV because even under the best conditions only 0.25%–1% of the inoculum is infectious (KNUDSON and TINSLEY 1974; VOLKMAN et al. 1976). Thus, it is perhaps not surprising that fusion, phagocytosis, and viropexis (adsorptive endocytosis) have all been described as mechanisms of BV entry.

One mechanism of entry that can be linked experimentally to infection is adsorptive endocytosis. This route of entry is used for infection of vertebrate cells by Semliki Forest virus, influenza, vesicular stomatitis virus, and others (HELENIUS and MARSH 1982). The model for infection via adsorptive endocytosis is illustrated in Fig. 6. Viruses entering by adsorptive or receptor-mediated endocytosis adsorb to the surface of the host cell and are taken into clathrin-coated pits which pinch off to become coated vesicles. Soon thereafter, the clathrin dissociates from the coated vesicles and two or more may fuse to become an endosome. The pH inside the endosome decreases, inducing the viral envelope and the endosomal membrane to fuse, resulting in the penetration of the viral nucleocapsid into the host cell cytoplasm. Viruses that functionally enter by

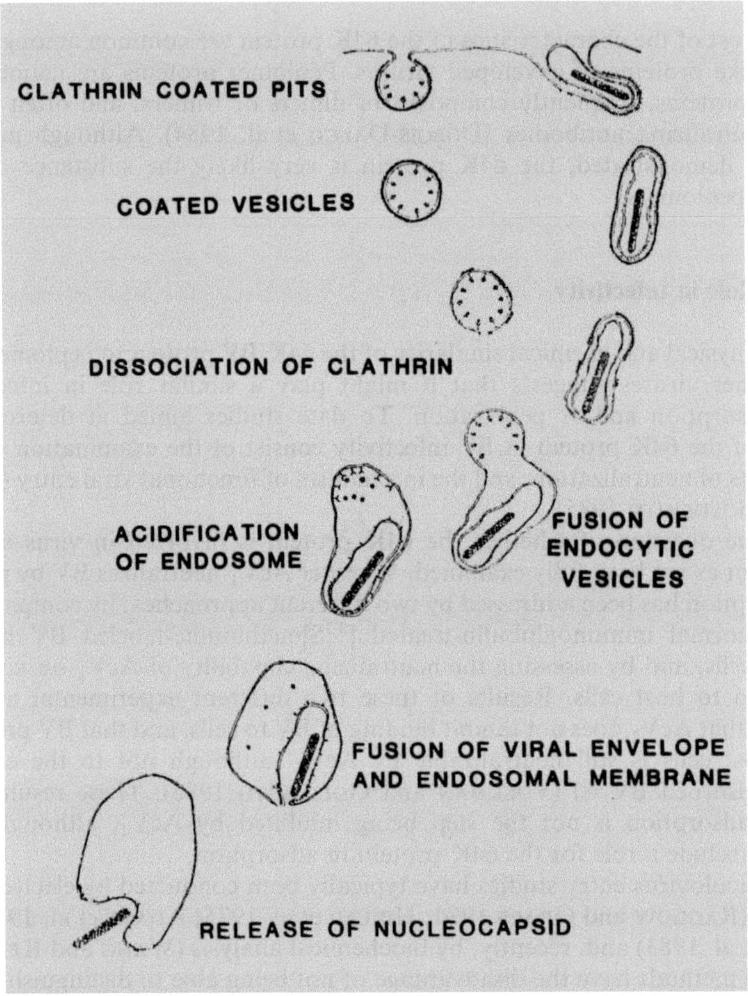

Fig. 6. Some viruses enter cells via an endocytocic pathway normally used to take up molecules involved in metabolism. Viruses and macromolecules bind to the cell surface, then are gathered into invaginations of the plasma membrane called "pits". These pits, coated with a protein called clathrin, pinch off to become coated vesicles. Soon after, the clathrin dissociates from the vesicles and two or more may fuse to become a larger vesicle, an endosome. The pH of the endosome decreases, presumably via an ATP-driven proton pump, and induces the viral envelope and the endosomal membrane to fuse, releasing the nucleocapsid into the cytoplasm

this pathway typically can be prevented from establishing infection by the presence of lipophilic amines. The lipophilic amines prevent the decrease of endosomal pH and thereby prevent fusion of the viral envelope and the endosomal membrane. The inhibitory effect is transient, however, because once the viral nucleocapsid is released into the cytoplasm, infection can proceed in the presence of lipophilic amines.

Experiments performed to examine the sensitivity of infection of BV- and AcV_1-treated BV (to antibody excess) to the lipophilic amines chloroquine and NH_4Cl showed that only BV is sensitive, and that its sensitivity is transient (VOLKMAN and GOLDSMITH 1985). Further it was shown that BV, but not AcV_1-treated BV, can hemolyze red cells (a standard test for viral fusion activity) at pH 5.8 or less, but not at pH 6.0–7.0. These results indicate that BV functionally enters by the endocytic pathway, and that AcV_1 neutralizes BV by interfering with its efficient entry via this pathway. It should be noted that the infectivity of PDV in IPLB-21 cells is also not affected by lipophilic amines, nor can it hemolyze red cells at an acidic pH. These results suggest that the 64K envelope protein of BV is functional in the penetration step of BV, and perhaps has a role in low-pH-dependent fusion.

3.5 Is the 64K Protein a Component of PDV?

The answer to the question of whether BV and PDV contain the same species of 64K protein has not been resolved. Studies of AcNPV structural proteins have yielded conflicting results. DOBOS and COCHRAN (1980), for example, reported that the major BV 66K protein does not have a 66K counterpart in PDV, and STILES and WOOD (1983) could not detect a glycosylated 64K BV protein counterpart in PDV. On the other hand, SMITH and SUMMERS (1978) reported that both BV and PDV had 64K proteins, and MARUNIAK (1979) and MARUNIAK and SUMMERS (1981) not only concurred but additionally reported that these proteins were both glycosylated and phosphorylated, respectively.

Immunological studies are similarly conflicting. Antiserum to PDV does not neutralize BV (VOLKMAN et al. 1976; ROBERTS 1983; yet it reacts with the 64K BV protein by Western blot analysis (VOLKMAN 1983). The finding that neither AcV_1 nor any of the nonneutralizing monoclonal antibodies that react with the 64K BV protein react with PDV suggests that BV and PDV do not share the same molecular species of 64K (HOHMANN and FAULKNER 1983). Perhaps germane to this point is the observation that incubation of BV for 60 min at room temperature with dilute alkaline saline, the solution normally used to liberate PDV from polyhedra, does not destroy the AcV_1-reactive epitope (unpublished results). This finding indicates that the lack of reaction of AcV_1 with PDV is not attributable to the standard polyhedra dissolution procedure.

Another indication that the 64K proteins of BV and PDV are different comes from studies aimed at determining the function of the 64K protein in BV, where it appears to play a role in entry. It is clear from these studies that BV and PDV do not functionally enter cells in the same way, and that impairing the activity of the 64K protein with AcV_1 causes BV to lose efficiency in infection, both in cell culture and in the hemocoels of larvae, to the extent that its infectivity is comparable to PDVs in these locations (VOLKMAN and GOLDSMITH 1985; KEDDIE and VOLKMAN 1985). At the same time, AcV_1 treatment of BV causes it to lose its ability to become inhibited by lipophilic amines and hemolyze red cells at acidic pHs. In other words, BV with AcV_1-inactivated 64K protein components behaves in a similar way to PDV in these instances.

4 Summary and Conclusions

The 64K protein is a major envelope component of AcNPV BV. It is an acidic phosphoglycoprotein present in the virus as monomers, dimers, trimers, and tetramers, and probably composes the viral peplomers. It is the target of neutralizing monoclonal antibody AcV_1 and plays a role in the penetration step of BV infection. AcV_1 neutralizes BV by inhibiting efficient entry of BV by adsorptive endocytosis. AcV_1-treated BV infects cell culture cells and insect larvae via the hemocoel route with an efficiency that is several orders of magnitude less than untreated BV. The infectivity of AcV_1-treated BV in these locations is approximately the same as that of PDV, which has very little, if any, 64K protein. The difference in 64K protein content between BV and PDV, therefore, can account for the great difference in infectivity of the two forms both in cell culture and in the hemocoel of host insects. Thus the high concentration of 64K protein in the BV envelope represents an important structural modification that enables BV to readily carry out its task in nature of establishing systemic infections.

Acknowledgments. This work was supported by the U.S. Department of Agriculture Competitive Research Grant 83-CRCR-1-1244, U.S. Department of Agriculture S135 Regional Research Funds, and by Federal Hatch Funds. I thank Andrew Keddie for drawing Figs. 1 and 6, Roberta Hess for electron microscopy and photographic assistance, and Academic Press for permission to use Fig. 4, which was originally published in *Virology* 139:295–302 (1984).

References

Adams JR, Goodwin RH, Wilcox TA (1977) Electron microscopic investigations on invasion and replication of insect baculoviruses *in vivo* and *in vitro*. Biol Cell 28:261–268

Bassemir J, Miltenburger HG, David P (1983) Morphogenesis of nuclear polyhedrosis virus from *Autographa californica* in a cell line from *Mamestria brassicae* (cabbage moth). Cell Tissue Res 228:587–595

Carstens EB, Tjia ST, Doerfler W (1979) Infection of *Spodoptera frugiperda cells* with *Autographa californica* nuclear polyhedrosis virus I. Synthesis of intracellular proteins after virus infection. Virology 99:386–398

Dobos P, Cochran MA (1980) Protein synthesis in cells infected by *Autographa californica* nuclear polyhedrosis virus (Ac-NPV): the effect of cytosine arabinoside. Virology 103:446–464

Dubois-Dalcq M, Holmes KV, Rentier B (1984) Assembly of enveloped RNA Viruses, Springer, Vienna New York

Faulkner P (1981) Baculovirus. In: Davidson EW (ed) Pathogenesis of invertebrate microbial diseases, Allenheld, Osmun, New Jersey

Goldstein NI, McIntosh AH (1980) Glycoproteins of nuclear polyhedrosis viruses. Arch Virol 64:119–126

Granados RR (1980) Infectivity and mode of action of baculoviruses. Biotechnol and Bioeng 22:1377–1405

Helenius A, Marsh M (1982) Endocytosis of enveloped animal viruses. In: Membrane recycling, CIBA Foundation Symposium 92, Pitman, London

Hirumi H, Hirumi K, McIntosh AH (1975) Morphogenesis of a nuclear polyhedrosis virus of the alfalfa looper in a continuous cabbage looper cell line. Ann NY Acad Sci 266:302–326

Hohmann AW, Faulkner P (1983) Monoclonal antibodies to baculovirus structural proteins: determination of specificities by Western blot analysis. Virology 125:434–444

Keddie BA, Volkman LE (1985) Infectivity difference between the two phenotypes of *Autographa californica* nuclear polyhedrosis virus: importance of the 64K envelope glycoprotein. J Gen Virol 66:1195–1200

Kelly DC, Lescott T (1981) Baculovirus replication: protein synthesis in *Spodoptera frugiperda* cells infected with *Trichoplusia ni* nuclear polyhedrosis virus. Microbiologica 4:35–57

Kelly DC, Lescott T (1983) Baculovirus replication: glycosylation of polypeptides synthesised in *Trichoplusia ni* nuclear polyhedrosis virus infected cells and the effect of tunicamycin. J Gen Virol 64:1915–1926

Kelly DC, Brown DA, Ayres MD, Allen CJ, Walker IO (1983) Properties of the major nucleocapsid protein of *Heliothis zea* singly enveloped nuclear polyhedrosis virus. J Gen Virol 104:399–408

Knudson DL, Tinsley TW (1974) Replication of a nuclear polyhedrosis virus in a continuous cell culture of *Spodoptera frugiperda*: purification, assay of infectivity, and growth characteristics of the virus. J Virol 14:934–944

Maruniak JE (1979) Biochemical characterization of baculovirus structural and infected TN-368 cell polypeptides, glycoproteins, and phosphoproteins. PhD thesis, University of Texas, Austin

Maruniak JE, Summers MD (1981) *Autographa californica* nuclear polyhedrosis virus phosphoproteins and synthesis of intracellular proteins after virus infection. Virology 109:25–34

Padhi SB (1985) Viral proteins for the identification of insect viruses. In: Maramorosch K, Sherman KE (eds) Viral insecticides for biological control, Academic, London

Raghow R, Grace TDC (1974) Studies on a nuclear polyhedrosis virus in *Bombyx mori* cells *in vitro* I. Multiplication kinetics and ultrastructural studies. J Ultrastr Res 47:384–399

Roberts PL (1983) Neutralization studies on the *Autographa californica* nuclear polyhedrosis virus. Arch Vriol 75:147–150

Singh SP, Gudauskas RT, Harper JD, Edwards J (1985) Two-dimensional gel electrophoresis of basic proteins of *Autographa californica* nuclear polyhedrosis virus. J Invertebr Pathol 45:249–251

Smith GE, Summers MD (1978) Analysis of baculovirus genomes with restriction endonucleases. Virology 89:517–527

Stiles B, Wood HA (1983) A study of the glycoproteins of *Autographa californica* nuclear polyhedrosis virus (AcNPV) Virology 131:230–241

Summers MD, Volkman LE (1976) Comparison of biophysical and morphological properties of occluded and extracellular nonoccluded baculovirus from *in vivo* and *in vitro* host systems. J Virol 17:962–972

Tanada Y, Hess RT (1984) The cytopathology of baculovirus infections in insects. In: Kung RC, Arai H. Insect ultrastructure, vol 2. Plenum, New York

Tweeten KA, Bulla LA, Consigli RA (1980) Characterization of an extremely basic protein derived from granulosis virus nucleocapsids. J Virol 33:866–876

Vaughn J, Dougherty EM (1985) The replication of baculoviruses. In: Maramorosch K, Sherman KE (eds). Viral insecticides for biological control, Academic, London

Vlak JM, Rohrmann GF (1985) The nature of polyhedrin. In: Viral insecticides for biological control, Academic, London

Volkman LE (1983) Occluded and budded *Autographa californica* nuclear polyhedrosis virus: immunological relatedness of structural proteins. J Virol 46:221–229

Volkman LE, Goldsmith PA (1984) Budded *Autographa californica* NPV 64K protein: further biochemical analysis and effects of postimmunoprecipitation sample preparation conditions. Virology 139:295–302

Volkman LE, Goldsmith PA (1985) Mechanism of neutralization of budded *Autographa californica* nuclear polyhedrosis virus by a monoclonal antibody: inhibition of entry by adsorptive endocytosis. Virology 143:185–195

Volkman LE, Summers MD (1975) Nuclear polyhedrosis virus detection: relative capabilities of clones developed from *Trichoplusia ni* ovarian cell line TN-368 to serve as indicator cells in a plaque assay. J Virol 16:1630–1637

Volkman LE, Summers MD (1977) *Autographa californica* nuclear polyhedrosis virus: comparative infectivity of the occluded, alkali-liberated and nonoccluded forms. J Invertebr Pathol 30:102–103

Volkman LE, Summers MD, Hsieh C-H (1976) Occluded and nonoccluded nuclear polyhedrosis virus grown in *Trichoplusia ni*: comparative neutralization, comparative infectivity, and *in vitro* growth studies. J Virol 19:820–832

Volkman LE, Goldsmith PA, Hess RT, Faulkner P (1984) Neutralization of budded *Autographa californica* NPV by a monoclonal antibody: identification of the target antigen. Virology 133:354–362

Volkman LE, Goldsmith PA, Hess RT (1986) Alternate pathway of entry of budded *Autographa californica* nuclear polyhedrosis virus: fusion at the plasma membrane. Virology 148:288–297

Wang H, Kelly DC (1985) Baculovirus replication: uptake of *Trichoplusia ni* nuclear polyhedrosis virus particles by insect cells. J Gen Virol 66:541–550

Wood HA (1980) *Autographa californica* nuclear polyhedrosis virus-induced proteins in tissue culture. Virology 102:21–27

Persistent and Productive Infections with the Hz-1 Baculovirus

H.A. Wood[1] and J.P. Burand[2]

1 Introduction

1.1 Nature of Persistent Virus Infections

Persistent virus infections are among the most interesting and complex host: parasite interactions found in nature. This dynamic coexistence has been described in all classes of living organisms. The common element in persistent infections is that host cells are able to carry out most processes normally and can continue to divide. Persistent virus infections may eventually kill the host but only after a long period of association.

Persistent infections are generally classified as latent, chronic, or slow. Latent infections describe conditions under which the virus is usually not detectable and intermittent acute symptoms may occur. During chronic infections, the virus is always detectable but the symptoms are usually inapparent. Slow infec-

[1] Boyce Thompson Institute for Plant Research at Cornell University, Tower Road, Ithaca, NY 14853 USA
[2] Department of Biology, University of Southern Colorado, Pueblo, CO 81001, USA

Current Topics in Microbiology and Immunology, Vol. 131

tions are characterized as having long incubation periods followed by a slowly progressive disease that is usually lethal. Often the varying manifestations which occur during persistent infections make them difficult to classify into one of these categories.

The host and virus properties which work in concert to establish persistent infections are varied. Persistence can occur when immune systems are altered or infections present the system with few or inaccessible antigens. Viral genomes may become integrated into the host chromosome with little or no expression. Replication of the virus may be altered through defective-interfering particles or temperature-sensitive mutations. Virus infections also may be nonpathogenic to a particular host.

The sophisticated nature of the interaction required to establish persistent infections suggests a close evolutionary involvement between the host and virus. From an evolutionary standpoint persistent infections have a selective advantage because survival of both the host and the pathogen is assured. Virus infection of a host for which no evolutionary coadaptation has occurred would be expected to elicit a mild to lethal disease.

Persistent virus infections of insects have usually been described as latent infections. The arboviruses, such as Togaviridae and Bunyaviridae, replicate in insect hosts with little or no deleterious effect yet often cause lethal diseases in vertebrates. Nodamura virus infection of the mosquito *Culex tritaeniorhynchus* is asymptomatic but is lethal to honey bees (*Apis mellifera*), greater wax moth (*Galleria mellonella*), and mice (BAILEY et al. 1975). Viruses in the polydnaviridae have established an obligate mutualistic relationship with their endoparasitic hymenopterous hosts. In lepidopterous hosts, these viruses interfere with host defense mechanisms against parasitism by the wasp. The ability to establish a persistent infection in one host species but not in another is common. Persistent infections arise through a combination of viral and host properties which work in an integrated fashion to establish a persistent interaction.

Unlike other persistent viruses of animals, the Hz-1 virus possesses properties that permit the establishment of a persistent interaction with all known host cells. It has a wide host range amongst lepidopterous cell cultures, and the establishment of productive and persistent infections can be easily manipulated. Therefore, the Hz-1 virus is an excellent model system for studying the establishment and maintenance of persistent infections of a single virus in several hosts. This unique situation offers a significant opportunity to investigate many of the basic principles concerning the general nature of persistent animal virus infections.

1.2 Hz-1 Virus: Discovery

In 1970, HINK and IGNOFFO reported establishing the IMC-Hz-1 tissue culture cell line from adult ovarian tissue of *Heliothis zea*. The purpose of establishing this cell line was to study the replication of the *H. zea* nuclear polyhedrosis virus (HzNPV). When the cells were inoculated with HzNPV, the tissue culture cells died and the culture fluids were lethal to *H. zea* larvae (IGNOFFO et al.

1971). However, no virus particles or occlusion bodies (typically found late in NPV infections) were detected, and after seven serial passages infectious virus replication was not detected.

In 1972, HINK reported a personal communication in which K.L. ROY observed long flexuous virus-like rod particles in IMC-Hz-1 cells. Based on this observation and his previous studies, Hink speculated that these rods may be latent viruses. He felt that a latent virus infection could explain why this cell line was not susceptible to other virus infections.

In further studies, VAIL et al. (1973) inoculated the IMC-Hz-1 cells with the *Autographa californica* (Ac) NPV. He reported that the cells became granular in appearance and ceased division. As with previous studies, no virus occlusion bodies were observed. Since this cell line appeared nonsusceptible to most insect viruses or gave unexpected cytopathic effects, the IMC-Hz-1 cell line was considered of interest only as a laboratory curiosity.

In 1978, GRANADOS and coworkers investigated the properties of these cells more closely. They observed that the cells often entered a crisis phase during which most cells were granulated, shriveled in appearance, and died. The remaining cells were used to reestablish the culture. By electron microscopy, they found enveloped, rod-shaped nucleocapsid structures which resembled long baculovirus particles in the IMC-Hz-1 cells. They transmitted this virus to *Porthetria dispar* (IPLB-65Z), *H. zea* (IPLB-1075), *Spodoptera frugiperda* (IPLB-SF-21), and *Trichoplusia ni* (TN-368) tissue culture cells. Virus occlusion bodies were never observed in any of the infected cells, which became granulated and died soon after inoculation. Through these studies it was established that the IMC-Hz-1 cells were persistently infected.

In retrospect, we now know that the IMC-Hz-1 cells were refractory to infections with the HzNPV and AcNPV viruses because of the persistent Hz-1 infections. The cytopathic effects observed resulted from the induction of productive Hz-1 replication and subsequent lysis of the cells. The crisis phase observed is not uncommon with persistently infected cells. The Hz-1 virus is a nonoccluded baculovirus (subgroup C of Baculoviridae) and does not share significant sequence homology with the HzNPV, AcNPV, or *Lymantria dispar* NPV occlusion body genes (BURAND and WOOD, unpublished observations).

2 Physical Properties

The Hz-1 virus population isolated from IMC-Hz-1 cells is a "wild" isolate and as such was considered to be genetically heterogenous. In an effort to isolate and characterize a genetically homogenous strain of virus using plaque purification procedures, two types of Hz-1 virus populations, standard and defective, were routinely isolated (BURAND et al. 1983a). The defective virus population always contains a low percentage of standard particles.

The standard Hz-1 virus particles contain a superhelical, circular, double-stranded DNA genome of 233 kb (HUANG et al. 1982; BURAND et al. 1983a). Restriction endonuclease fragment patterns (Fig. 1) show no detectable minor molar fragments. The standard virus particles are bacilliform, measuring $414 \pm 30 \times 80 \pm 3$ nm (Fig. 2A). The Hz-1 virus particle length and genome are approximately twice that of subgroup A baculoviruses, suggesting similar packaging of the genome.

Twenty-eight structural proteins ranging in molecular weight from 153,000 to 14,000 have been identified in standard virus particles (Fig. 3) (BURAND et al. 1983b). Fourteen of these proteins are glycosylated based on labeling with [^{3}H]mannose and *N*-[^{3}H]acetylglucosamine. In the presence of tunicamycin, incorporation of labeled sugars was not detected, and there was a 97% reduction in virus titers. Therefore, glycosylation is considered to be via *N*-glycosidic bonds and is required for virus infectivity.

The defective Hz-1 virus population contains a low percentage of standard particles which are probably required for the replication of the defective particles. The defective particles are heterogenous in length (370 ± 76 nm) (Fig. 2B) and contain genomes with deletions of up to 100 kb. The heterogenous nature of the genomic deletions are apparent in Fig. 1. Many minor molar fragments are apparent along with a background of the standard viral genome. As determined by thermal denaturation studies, the defective viral population DNAs have a major and minor component with 43 and 68% G+C (HUANG et al. 1982). Due to the large size of the Hz-1 virus genome, a complete physical map is yet to be developed. Preliminary studies have shown, however, that the defective particles most likely arise from deletions in a single region of the standard virus genome (WOOD and BURAND, unpublished). The defective virus particles contain the same structural proteins detected in standard virus particles.

3 Hz-1 Virus Persistence

3.1 Establishment of Persistent Infections

Inoculation of tissue culture cells with standard virus particles initially results in lysis of all the cells within 14 h post-inoculation (pi). However, following undiluted serial passage of the standard virus, a high level of defective particles are generated and some of the cells survive. These surviving cells are persistently infected (BURAND et al. 1983a). In a yet to be determined manner, the defective particles interfere with productive replication of the standard particles. The persistently infected cells have a morphologically normal appearance and survive inoculation even with standard Hz-1 virus particles.

Southern blots of DNA purified from persistently infected cells probed with viral DNA appeared similar to the genomic composition of the defective virus population (BURAND et al. 1983a). With one of the persistently infected *T. ni* cell cultures, approximately 500 viral genome copies per cell were detected. Yet the cells exhibited no cytopathic manifestations.

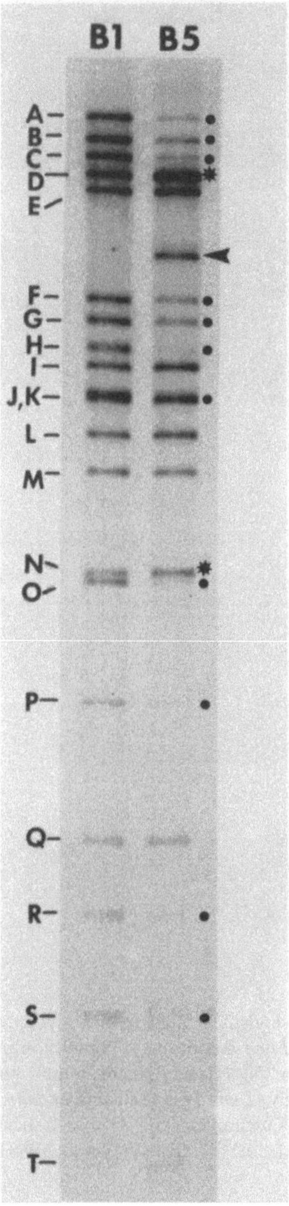

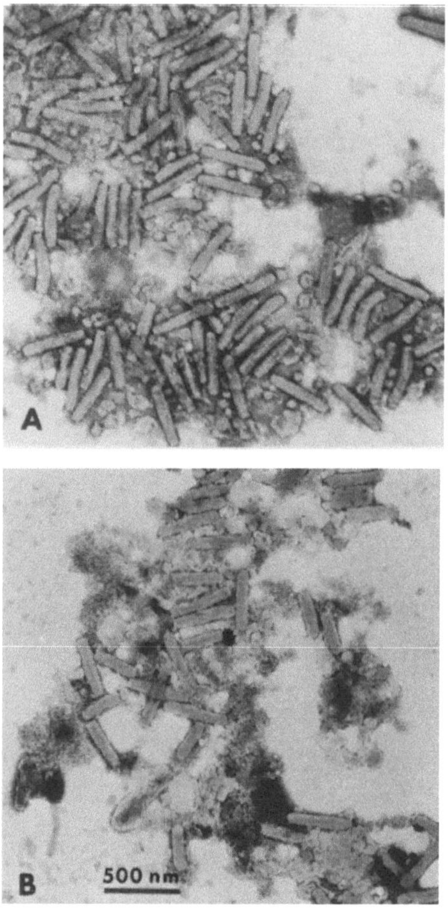

Fig. 2A, B. Electron micrographs of standard **A** and defective **B** virus particles. (BURAND et al. 1983a)

Fig. 1. *Eco*RI restriction enzyme profile of standard (*B1*) and defective (*B5*) DNA. The fragments are identified by *letters*. *Asterisks and small circles* indicate increased and decreased fragment concentrations, respectively, in the defective particle DNA. The *arrow* indicates an additional fragment. (BURAND and WOOD 1986)

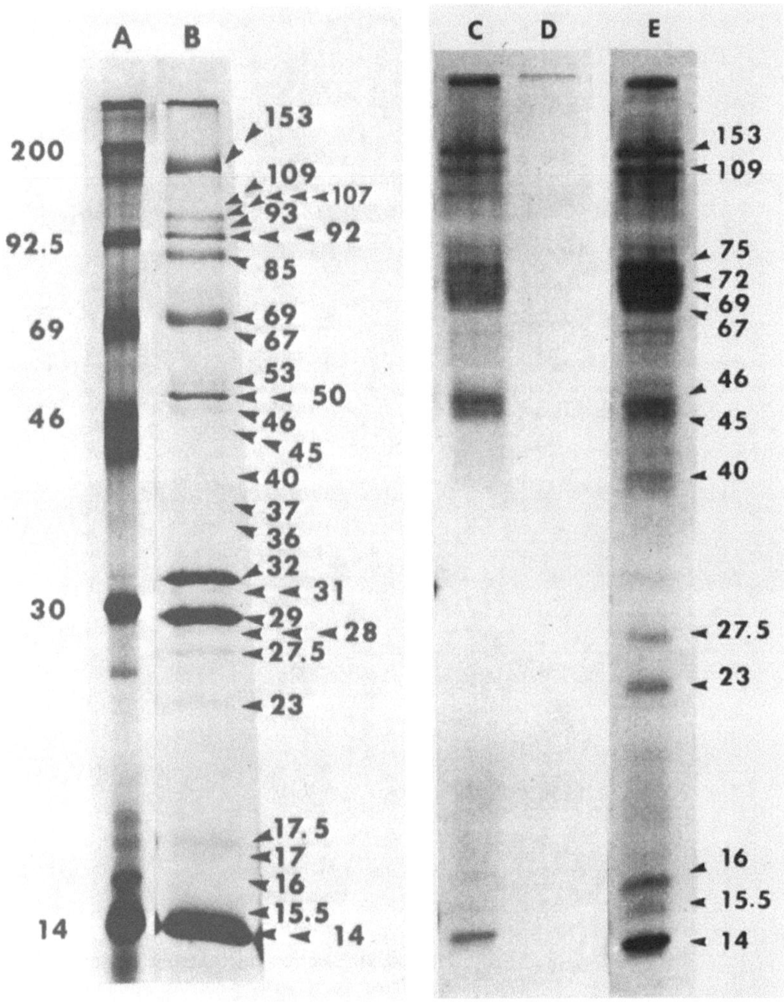

Fig. 3. Autoradiograms of PAGE fractionated Hz-1 virus structural and glycoproteins. *Lane A* contains molecular weight standards. *Lane B* contains [^{35}S]methionine-labeled viral proteins. *Lanes C* and *E* contain N[^{3}H]acetylglucosamine- and [^{3}H]mannose-labeled viral glycoproteins, respectively. *Lane D* is the same as *lane C* but labeling was done in the presence of 10 µg tunicamycin/ml. *Arrows* indicate viral proteins and their apparent molecular weights ($\times 10^{-3}$). (BURAND et al. 1983b)

3.2 Host Range

The reported host range of the Hz-1 virus includes cell lines from the following lepidopterons: *T. ni* (TN-368), *Spodoptera frugiperda* (IPLB-SF-212), *H. zea* (IPLB-1075), *Mamestra brassicae*, and *Porthetria dispar* (IPLB-65Z) (GRANADOS et al. 1978; RALSTON et al. 1981; and KELLY et al. 1981). In addition we have identified the *Lymantria dispar* (LD 252Y) and *Heliothis virescens* (BCIRL-HB-AM1) cell lines a susceptible to Hz-1 infections. Only the *Estigmene acrea* (BTI-EAA) (GRANADOS et al. 1978) and *Manduca sexta* (RALSTON et al. 1981) cell cultures have been judged to be nonhosts or semipermissive hosts based on infectious titers at 2 days pi and healthy appearance.

Persistent Hz-1 virus infections have been established in *T. ni* (TN-368) cells (RALSTON et al. 1981). In addition, we have established persistent infections in *S. frugiperda* cells (unpublished). RALSTON et al. (1981) reported that two cloned cell lines derived from TN-368 could support productive replication of the Hz-1 virus but persistently infected cultures could not be established.

Based on symptom expression, infection of larvae with the Hz-1 virus has not been documented. The larvae tested include *Estigmene acrea*, *H. zea*, *H. armigera*, *S. frugiperda* and *S. littoralis* (GRANADOS et al. 1978 and KELLY et al. 1981). However, the report by IGNOFFO et al. (1971) indicates lethal infections of *H. zea* larvae.

It should be noted that all of the host range studies mentioned above were performed with virus populations containing defective particles which could have established persistent infections; such infections could be overlooked if lethality was used as the criterium for host status. For instance, RALSTON et al. (1981) observed in some experiments that more than 50% of the *T. ni* cells (the most sensitive indicator cells described) became persistently infected following inoculations with defective-virus samples. Therefore, a determination of the host range of Hz-1 virus should be done only with preparations of standard virus particles (eliciting only productive replication in cell cultures) or further testing must be conducted to detect persistent infections.

3.3 Nature of Persistent Infections

Persistently infected IMC-Hz-1 and TN-368 cells continuously release low levels of infectious virus into the medium. The persistently infected TN-368 cells appear healthy yet grow slightly slower. At irregular intervals the cultures undergo a crisis period in which a majority of the cells lyse. The cytopathic effects mimic productive replication with concomitant increases in virus titers.

The persistently infected IMC-Hz-1 cells are resistant to infection with all baculoviruses tested (KELLY et al. 1981: GRANADOS, personal communication). However, KELLY et al. (1981) reported that inoculation with Hz-1 (superinfection), or NPVs of *T. ni*, *S. frugiperda*, *S. littoralis* and *H. armigera* induced productive replication of the Hz-1 virus. This induction also occurred with heat- or UV light-inactivated *T. ni* NPV. Productive induction was not observed following inoculation with iridescent virus type 22 or inoculation with *T. ni* NPV in the presence of actinomycin D, azetidine, canavanine or cycloheximide. Induction by *T. ni* NPV could proceed in the presence of hydroxyurea and cytosine arabinoside.

In contrast, inoculation of persistently infected TN-368 cells with the Hz-1 virus has no effect on the cells (RALSTON et al. 1981). In fact, we consider survival following a second Hz-1 virus inoculation to be indicative of the establishment of a persistent infection in *T. ni* and *S. frugiperda* cells. Also, inoculations of persistently infected TN-368 cells with the AcNPV results in AcNPV replication *not* Hz-1 activation as with the persistently infected IMC-Hz-1 cells (RALSTON et al. 1981).

The resistance to heterologous baculovirus infections of IMC-Hz-1 cells but susceptibility of persistently infected TN-368 cells is probably a property of cellular determinants. Another consideration is that the type of persistent interaction established with these two cell types may be different. The induction of Hz-1 productive replication via superinfection is an unusual response of persistently infected cell cultures.

4 Replication

4.1 Productive Replication

Infection with standard Hz-1 particles results in lysis of all host cells tested. Inoculations with samples containing both defective and standard virus particles usually result in the survival of less than 10% of the cells, and these survivors are persistently infected. RALSTON et al. (1981) observed, however, that persistent infections of the C10 and C13 clones of TN-368 cells could not be established. This observation emphasizes the integrated relationship which must be established between a virus and host to initiate a persistent infection.

The defective particles not only mediate the establishment of persistent infections but also interfere with the infection and/or replication of standard particles (BURAND and WOOD 1986). Inoculation of TN-368 cells first with standard and then with defective virus populations resulted in a 74% reduction in titer as compared with standard virus alone. The reverse inoculation order resulted in an 84% reduction. Therefore, the defective particles are defective interfering (DI) particles.

The growth curve of Hz-1 in TN-368 cells inoculated with standard virus particles is shown in Fig. 4 (BURAND et al. 1983). Infectious virus associated with cells was first detected at 8 h pi and reached a maximum titer at 15 h pi. The increase in extracellular virus titers was not detected until 12 h pi and reached a maximum at about 20 h pi. The time shift in extracellular and cell-associated virus titers occurs because maturation of infectious particles within the cells occurs prior to lysis. RALSTON et al. (1981) obtained similar data for

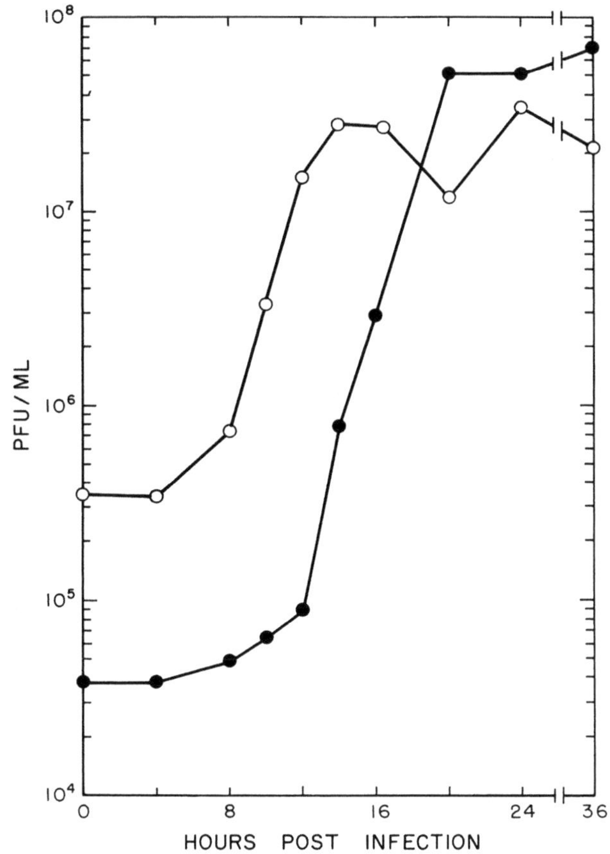

Fig. 4. Growth curve of standard Hz-1 virus in TN-368 cells. *Open circles* are infectivity of cell-associated virus. *Closed circles* indicate infectivity of extracellular virus. (BURAND et al. 1983b)

extracellular virus titers following DI particles-containing inoculations of TN-368 cells. However, they observed an almost four fold decrease in titers from 24 to 36 h pi. This decrease could have resulted from aggregation of virions or adsorption of particles to cellular debris.

4.2 Electron Microscopy

R.R. GRANADOS (manuscript in preparation) has recently completed an ultrastructural examination of TN-368 cells inoculated with Hz-1 virus. He observed that virus entry occurred by fusion of the viral envelope with the cell plasma membrane and insertion of the nucleocapsid into the cell cytoplasm. However, entry of enveloped nucleocapsids by viroplexis could not be excluded since virions within phagocytic vacuoles were occasionally observed. The nucleocapsids became attached to the nuclear pore complex, and the viral genome was released through the nuclear pore into the nucleus.

The first morphogenic changes which he observed were at 8 h pi. They consisted of an enlargement of the nucleus and margination of the cell chromatin. Within the virogenic stroma, viral morphogenesis commenced with the formation of electron-dense cores (nucleocapsids) within discrete accumulations of open-ended, membrane structures. By 16 h pi openings in the nuclear envelope allowed the enveloped viral particles to enter the cytoplasm. The virions accumulated along the plasma membrane but budding through the membrane was not observed. During the final stages of the infection process (18–24 h pi), lysis of the plasma membrane released the virions and cellular organelles into the growth medium.

4.3 Induced Productive Replication in IMC-Hz-1 Cells

Kelly et al. (1981) inoculated IMC-Hz-1 cells with *H. armigera* NPV as a means of inducing productive replication of the Hz-1 virus. By 5 h pi a condensation of the cell chromatin was observed. Viral nucleocapsids were observed in the nucleus at 12 h pi. These particles were closely associated with a virogenic stroma. By 15 h pi all cells showed cytopathic effects and the virions were observed to become membrane bound within the nucleus. A disintegration of the nuclear membrane was observed and was followed by a release of the particles through lysis of the cell membrane.

4.4 Viral DNA Replication

Hz-1 viral DNA replication is first detected at 4 h pi in TN-368 cells (Burand and Wood 1986). The onset and rates of DNA replication were the same following standard and DI particle inoculations. This would be expected since even with DI particle inoculations the majority of the cells are subject to productive replication and lysis.

The amount of total DNA replication in infected cells as measured by [^{3}H]thymidine incorporations agrees with the amount of viral DNA detected by DNA: DNA dot blot hybridization techniques. Therefore, as with other baculoviruses (Knudson and Tinsley 1974), it appears that Hz-1 infection interrupts host DNA replication.

4.5 Viral-Induced Proteins

4.5.1 Standard Virus Inoculations

Pulse labeling studies using [^{35}S]methionine or [^{14}C]amino acids have been used to identify 37 Hz-1 viral-induced proteins (VIP) during the replication cycle in TN-368 cells (Burand et al. 1983b). As shown in Fig. 5, prior to the detection of viral DNA replication (early phase), only a 75K VIP was detected. During an intermediate stage, the time prior to intracellular viral maturation,

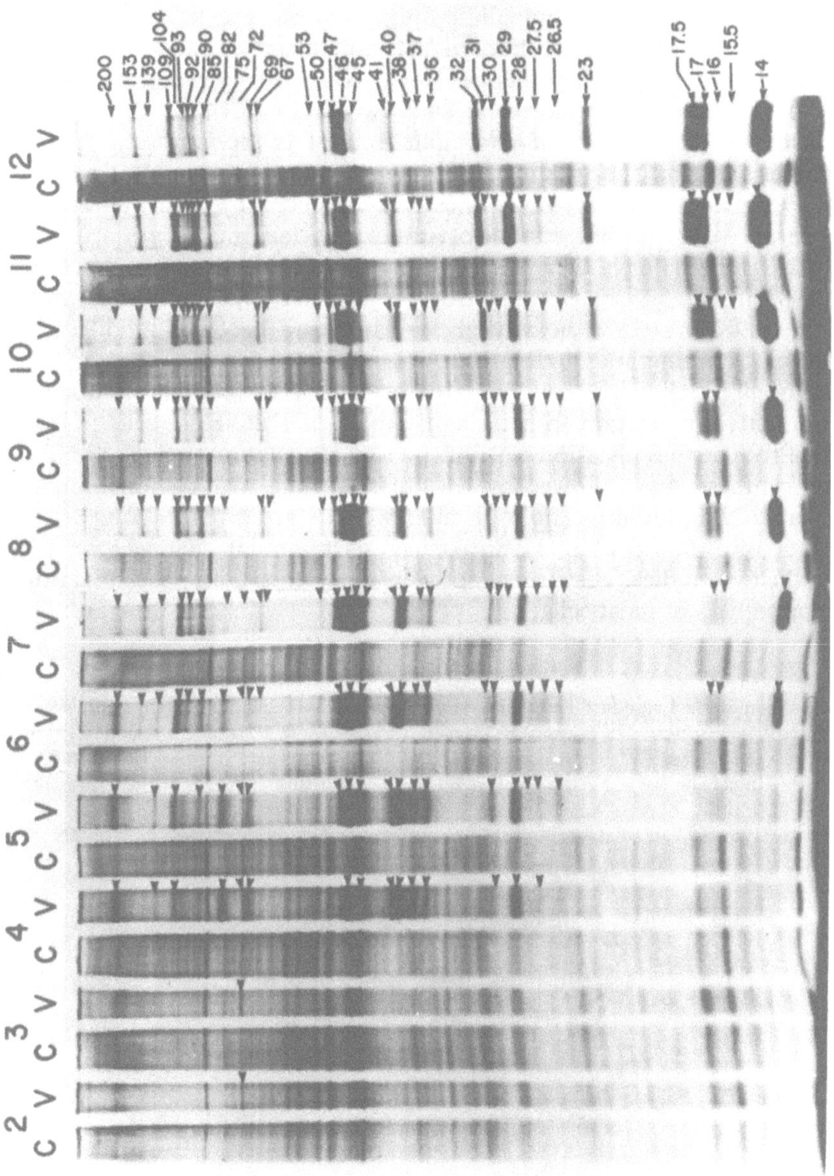

Fig. 5. Autoradiogram of intracellular polypeptides pulse labeled for 1 h with [^{35}S]methionine following inoculation of TN-368 cells with standard Hz-1 virus particles. Mock-inoculated (*C*) and inoculated (*V*) cell samples were analyzed by PAGE. The apparent molecular weights ($\times 10^{-3}$) of the viral-induced proteins are shown *on the right*. (BURAND et al. 1983b)

an additional 29 VIP were first detected. There is dramatic turning off of host cell protein synthesis in the intermediate stage. During the late stage, virus maturation, and cell lysis, seven additional VIP were observed.

As expected, the early 75K protein is not a Hz-1 virus structural protein, and all of the late-stage VIP have molecular weights equivalent to those of structural proteins (Figs. 3, 5). Of particular interest is the temporal control of VIP synthesis during the intermediate stage. There is a temporal control of the 72K, 75K, and 82K protein synthesis as evidenced by their appearance and disappearance during this period. Post-transcriptional modifications of the 109K and 82K proteins have been identified by pulse chase experiments (BURAND et al. 1983b). Recent experiments (BURAND and WOOD, unpublished) indicate that the 109K protein is heavily glycosylated and becomes modified by deglycosylation, resulting in a 107K protein.

Viral-induced protein synthesis has also been studied following induction of Hz-1 productive replication in persistently infected IMC-Hz-1 cells (KELLY et al. 1981). Productive replication was induced by inoculation with *T. ni* NPV, and protein synthesis was followed by 1 h pulse labeling with [^{35}S]methionine at 5, 9, 15, and 22 h. Although no molecular weights are given, a comparison with the data of BURAND et al. (1983b) suggests a similar sequence of VIP synthesis in the two studies. Very few bands were detected at 22 h post-induction, probably due to lysis of most cells.

Table 1. Comparison of Hz-1 standard and defective virus-specific intracellular protein synthesis

Mol. wt (10^{-3})	Detected in DI Infected cells (h pi)	Alteration in synthesis[a]	Structural protein[b]	Structural glycoprotein[c]
109	4–10	–	+	+
107	4–10	+	+	
93	8–12	–	+	
92	10–12	–	+	
90	10–12	–		
85	10–12	–	+	
75	8–12	+		+
69	6–12	–	+	+
50	7–12	–	+	
41	5–12	+		
40	3–9	–	+	+
37	5–12	+	+	
27.5	4–10	–	+	+
23	8–12	–	+	+
17.5	6–12	–	+	
17	7–12	–	+	
14.2	9–12	+		
14	5–12	–	+	+

[a] + and – indicate increase or decrease in relative rates of synthesis during DI virus replication compared with the rates of synthesis during standard virus replication

[b] BURAND et al. (1983a)

[c] From BURAND and WOOD (1986)

4.5.2 Defective Interfering Particle Inoculations

Inoculation of TN-368 cells with DI particle preparations of the Hz-1 virus result in the synthesis of the same 37 viral-induced proteins as found following standard virus inoculations (BURAND and WOOD 1986). These proteins were also detected at approximately the same time pi with both inocula. However, as indicated in Table 1, there are significant differences in the rates of synthesis of several viral-induced proteins. The regulatory changes which occur in the presence of DI particles include a reduction in the synthesis of nine nonstructural and five of the nine major virus structural proteins. The rate of synthesis of the 37K structural protein as well as four nonstructural proteins increases. In addition, the glycosylation of the 107K protein (a 109K glycoprotein) seems to be interrupted by the DI particles.

5 Conclusions

The significance of the Hz-1 virus lies in the fact that it is one of the most novel and perhaps most amenable systems for advancing our understanding of the basis of persistent animal virus infections. As we have noted, Hz-1 virus can give rise to both productive and persistent replication in a number of different cell types. This is very unusual and provides a unique opportunity to study the establishment and maintenance of persistence of a single virus in several different hosts.

It is conceivable that persistence may be mediated by different mechanisms in different hosts. For instance, the IMC-Hz-1 cells are refractory to infection by all NPVs tested, while persistently infected *T. ni* cells are susceptible. This suggests that the resistant nature of the IMC-Hz-1 cells is at least partially a property of cellular determinants. In addition, IMC-Hz-1 cells but not the persistently infected *T. ni* cells are susceptible to superinfection with the Hz-1 virus. In fact, uninfected *T. ni* and *S. frugiperda* cells can be eliminated from persistently infected cell cultures by introduction of the virus.

To date there are no firm data concerning the number of cells which carry and shed virus in a persistently infected culture. Based on electron microscopy, less than 25% of the IMC-Hz-1 cells contain virus particles, and the amount of infectious virus in the cell media varies during the transfer of cells. In contrast to this, superinfection can result in productive replication in all IMC-Hz-1 cells. Therefore, it would appear that at any point in time only a small number of cells are producing infectious virions but that most cells carry the viral genome, either as an episome or integrated into the host chromosome.

Among the viruses classified in the Baculoviridae, Hz-1 virus infections are uncharacteristic with respect to host response. During productive replication, cell lysis occurs after 12 h pi. With other baculoviruses, lysis of cells usually occurs after 36 h pi and is presumably associated with an inability to maintain processes required for membrane integrity. Since Hz-1 virions do not bud through the plasma membrane, we suspect that the virus codes for a lytic factor which is required for virus release. In addition, even at low multiplicities of infection with the Hz-1 virus, there is a shutdown of host protein synthesis early in the infection cycle. This would suggest a positive control over the protein synthesis machinery. With other baculoviruses, this usually occurs late in the infection cycle and is suspected to result from competition of viral mRNAs for ribosomal binding sites. Therefore, in addition to the ability to establish persistent infections, the lack of occlusion body formation and increased genomic size, Hz-1 virus also differs significantly from other baculoviruses in its replication.

At the present time the Hz-1 virus is the best model available for the study of persistent insect virus infections. Because of its unique properties this virus may also be useful in answering many of the basic questions posed concerning persistant animal virus infections in general. As a biological agent for insect control (proposed for most lepidopterous baculoviruses), the Hz-1 virus may have only limited potential because it lacks an occluded form required for stability in the external environment. However, the Hz-1 virus carries important genetic information for early cell lysis and the shutoff of host cell protein synthesis. These genetic elements are prime candidates for gene transfer into occluded baculoviruses. Shutting off host protein synthesis with a genetically engineered baculovirus would allow for more efficient virus replication and quicker death of target insects. If expression of the Hz-1 virus lytic factor were under the control of a late gene of an occluded virus, the death rate of infected organisms would also be increased. In conclusion, future investigations with the Hz-1 virus may be useful in solving problems which are both basic and practical in scope.

Acknowledgments. We wish to thank Dr. R.R. Granados for his unpublished research results.

References

Bailey L, Newman JFE, Porterfield JS (1975) The multiplication of Nodamura virus in insect and mammalian cell cultures. J Gen Virol 26:15

Burand JP, Wood HA (1986) Intracellular protein synthesis during standard and defective Hz-1 virus replication. J Gen Virol 67:167

Burand JP, Wood HA, Summers MD (1983a) Defective particles from a persistent baculovirus infection in *Trichoplusia ni* tissue culture cells. J Gen Virol 64:391

Burand JP, Stiles B, Wood HA (1983b) Structural and intracellular proteins of the nonoccluded baculovirus Hz-1. J Virol 46:137

Granados RR, Nguyen T, Cato B (1978) An insect cell line persistently infected with a baculovirus-like particle. Intervirol 10:309

Hink WF (1972) Insect tissue culture. Adv Appl Microbiol 15:157

Hink WF, Ignoffo CM (1970) Establishment of a new cell line (IMC-Hz-1) from ovaries of the cotton bollworm moth, *Heliothis zea* (Boddie). Exp Cell Res 60:307

Huang Y, Hedberg J, Kawanishi CY (1982) Characterization of the DNA of a nonoccluded baculovirus, Hz-IV. J Virol 43:174

Ignoffo CM, Shapiro M, Hink WF (1971) Replication and serial passage of infectious *Heliothis* nuclear polyhedrosis virus in an established line of *Heliothis zea* cells. J Invertebr Pathol 18:131

Kelly DC, Lescott T, Ayres MD, Carey D, Coutts A, Harrap KA (1981) Induction of a nonoccluded baculovirus persistently infecting *Heliothis zea* cells by *Heliothis armigera* and *Trichoplusia ni* nuclear polyhedrosis viruses. Virol 112:174

Knudson DL, Tinsley TW (1974) Replication of a nuclear polyhedrosis virus in a continuous cell culture of *Spodoptera frugiperda*: purification, assay of infectivity, and growth characteristics of the virus. J Virol 14:934

Ralston AL, Huang Y, Kawanishi CY (1981) Cell culture studies with the IMC-Hz-1 nonoccluded virus. Virol 115:33

Vail PV, Jay DL, Hink WF (1973) Replication and infectivity of the nuclear polyhedrosis virus of the alfalfa looper, *Autographa californica*, produced in cells grown *in vitro*. J Invertebr Pathol 22:231

The Primary Structure of Baculovirus Inclusion Body Proteins. Evolution and Structure-Function Aspects

E.A. KOZLOV, T.L. LEVITINA, and N.M. GUSAK

1 Introduction

Interesting information about the relationships between the primary structure and physicochemical properties of the inclusion body protein can be obtained from studies on amino acid sequences. This approach allows a better understanding of the relationship between structure and function and the role of the protein in virus replication and virus–cell interactions. It is known that the inclusion body protein has highly specialized properties that allow it to form a protective crystal around the virions. However, there are several communications suggesting other functions of the baculovirus polyhedrin during infection: antitumor activity, hemagglutinating activity, induction of the nonspecific factor that inhibits macrophage migration, and ability to interact with DNA. Polyhedrin molecules in vivo can possibly not only interact with themselves (protein–protein interactions) but also with nucleic acids (protein–nucleic acid interactions) and membranes (protein–membrane interactions). The ability to interact with a wide scale of biopolymers is reflected in the character of the amino acid residue distribution along the polypeptide chain.

An attempt has been made to analyse theoretically the peculiarities of the primary structure of polyhedrin. Here, we would like to present a hypothesis

Institute of Molecular Biology and Genetics, Academy of Sciences of the Ukrainian SSR, Zabalotnogo 24, Kiev-143, USSR

Current Topics in Microbiology and Immunology, Vol. 131

on the structure-function organization of the polyhedrin polypeptide chain based on experimental data obtained from the study of the primary structure, the physicochemical properties of the protein and its fragments, and the theoretical analysis of the amino acid sequence.

The comparative studies of several polyhedrin and granulin primary structures will elucidate the phylogenetic relationships of baculoviruses. Physicochemical properties and amino acid sequences of *Bombyx mori, Porthetria dispar, Galleria mellonella, Agrotis segetum*, and *Mamestra brassicae* nuclear polyhedrosis virus (NPV) polyhedrins and *Agrotis segetum* granulosis virus (GV) granulin were investigated in our laboratory.

2 Primary Structure

A number of publications on the physicochemical properties of baculovirus inclusion body proteins (CIBULSKY et al. 1977; CRAWFORD and KALMAKOFF 1977; CROOK 1981; HARRAP et al. 1977; KOZLOV et al. 1975b, 1978; PADHI and CHASE 1976; ROHRMANN 1977; ROHRMANN et al. 1979, 1981; SCHARNHORST and WEAVER 1980; SUMMERS and SMITH 1975, 1976) seem to point to a high degree of homology in the primary structures of different nuclear polyhedrosis virus (NPV) polyhedrins. Proceeding from this premise we decided to determine the amino acid sequences of homologous proteins. The primary structure of the standard protein was established by the study of overlapping peptides and fragments obtained by enzymatic and chemical cleavage of the protein. The homologous proteins were cleaved by trypsin only. By comparing their structure with the sequence of the standard protein the tryptic peptides were localized on the polypeptide chain. Certain difficulties arose by taking this approach. When appreciable differences occurred between the tryptic peptides obtained and the corresponding peptide of the standard protein, peptide localization remained ambiguous. Moreover, when some short tryptic peptides could not be found it was not known whether they were lost during purification or whether there was a deletion in the gene of the investigated protein.

The *B. mori* NPV polyhedrin was chosen as a standard protein. This protein, obtained by dissolving polyhedra in 67% acetic acid (KOZLOV et al. 1975b), and its maleyl and carboxymethyl derivatives were cleaved by trypsin, chymotrypsin, and cyanogen bromide (LEVITINA et al. 1976; RADAVSKI et al. 1976). The peptides and fragments were fractionated by gel filtration, ion-exchange chromatography, paper electrophoresis, and chromatography (GUSAK et al. 1981, 1985; KOZLOV et al. 1973; LEVITINA et al. 1976, 1981; RADAVSKI et al. 1976). The amino acid sequence was determined by the manual Edman degradation procedures (GUSAK et al. 1979) and with the help of carboxypeptidases (KOZLOV et al. 1975b; LEVITINA et al. 1976; RADAVSKI et al. 1973). The data obtained made it possible to reconstruct the complete amino acid sequence of the polypeptide chain of *B. mori* NPV polyhedrin, as shown in Fig. 1.

The tryptic peptides of *P. dispar* (LEVITINA et al. 1981), *G. mellonella* (GUSAK et al. 1981; SIDOROVA et al. 1976) *A. segetum* (GUSAK et al. 1985), and *M. brassi-*

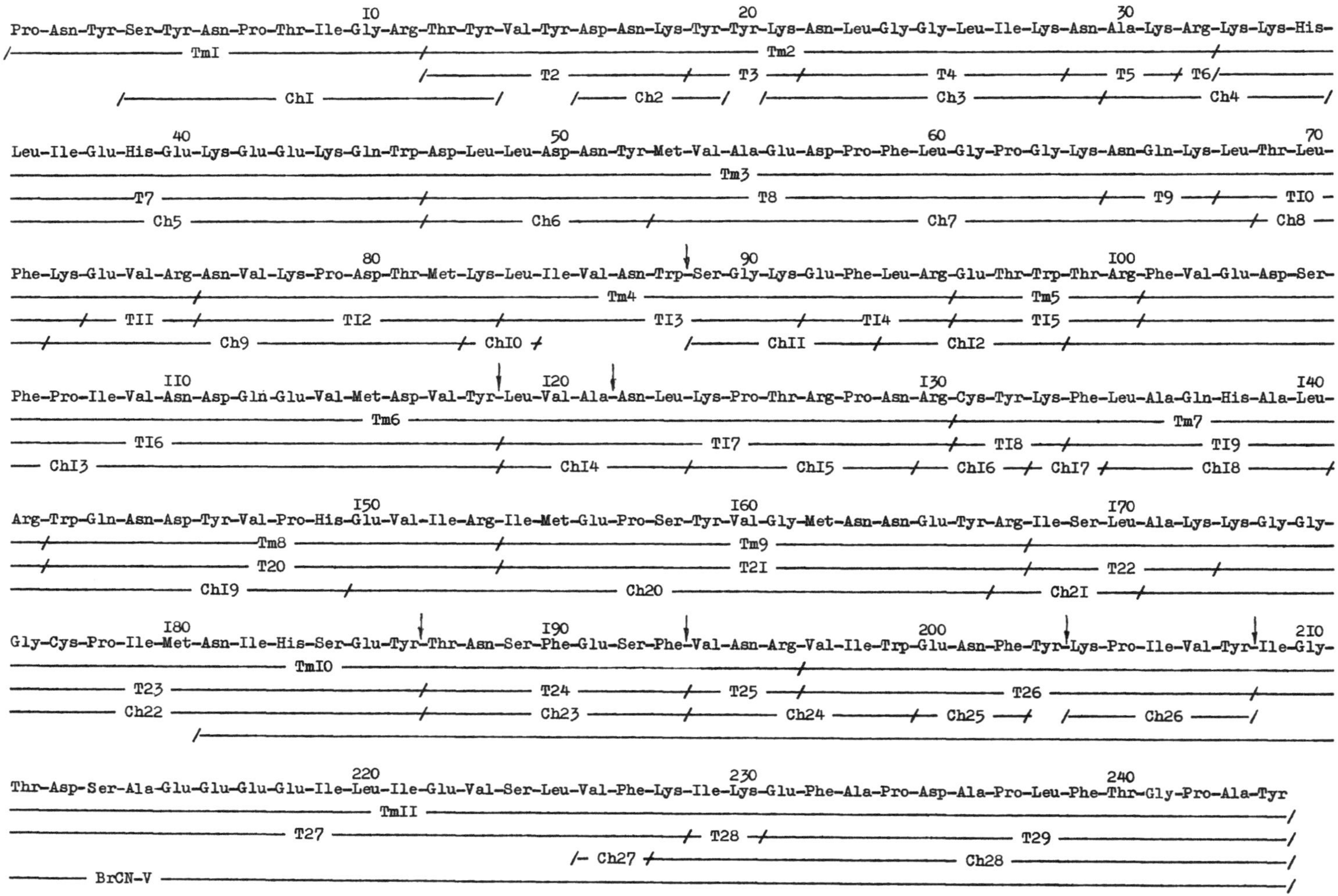

Fig. 1. Reconstruction pattern of *B. mori* NPV polyhedrin polypeptide chain. *T*, tryptic peptides; *Ch*, chymotryptic peptides; *Tm*, tryptic fragments of the maleylated protein. *Arrows* show the bonds that are cleaved by IB protease

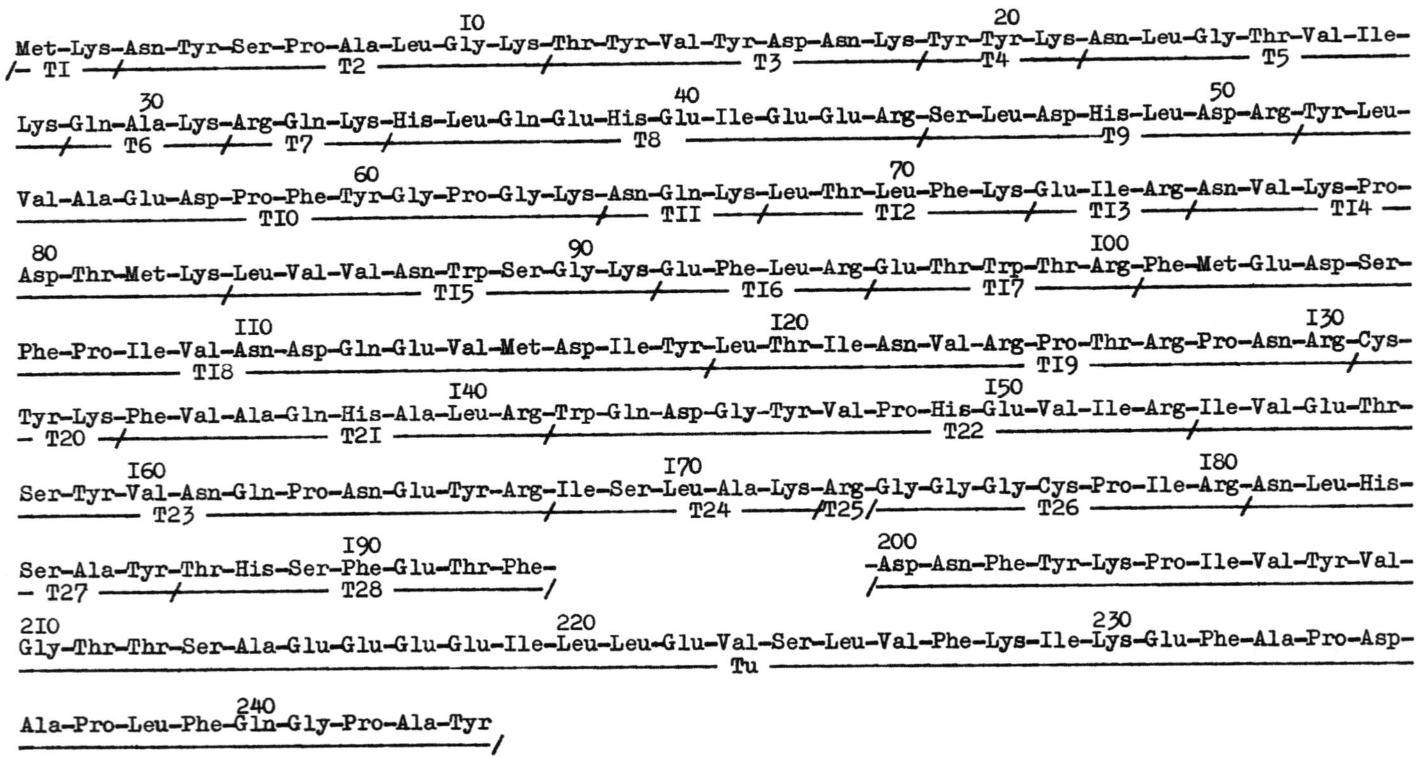

Fig. 2. Reconstruction pattern of *P. dispar* NPV polyhedrin polypeptide chain. The amino acid sequence numeration here and on Figs. 3–5 is in agreement with *B. mori* NPV polyhedrin sequence. The *gap* in 194–199 sequence infers that the peptides corresponding to that of *B. mori* NPV polyhedrin have not yet been found

10 20
Pro-Asn-Tyr-Ser-Tyr-(Arg,Thr,Pro,Gly,Ile)-Arg-Thr-Tyr-Val-Tyr-Asp-Asn-Lys-Tyr-Tyr-Lys-Asn-Leu-Gly-Ala-
/— T1 —/— T2 —/— T3 —/— T4 —

30 40 50
Val-Ile-Lys-Asn-Ala-Lys-Arg-Lys-Lys-(Lys,His_2,Asx_4,Glx_5,Gly,Val,Ile,Leu_3)-Leu-Val-Ala-Glu-Asp-Pro-Phe-
—/— T5 —/ T6/– T7 —/— T8 —/— T9 —

60 70 80
Leu-(Pro,Gly_2)-Lys-Asn-Gln-Lys-Leu-Thr-Leu-Phe-Lys-Glu-Ile-Arg-Asn-Val-Lys-Pro-Asp-Thr-Met-Lys-Leu-Val-
—/— T10 —/— T11 —/— T12 —/— T13 —/—

90 100 110
Val-Asn-Trp-Lys-Gly-Lys-Glu-Phe-Tyr-Arg-Glu-Thr-Trp-Thr-Arg-Phe-Met-Glu-Asp-Ser-Phe-Pro-Ile-Val-Asn-
– T14 —/ T15 —/— T16 —/— T17 —/— T18 —

120 130
Asp-Gln-Glu-Val-Met-Asp-Val-Phe-Leu-Val-Ala— Asn-Leu —Lys-Pro-Thr-Arg-Pro-Asn-Arg-Cys-Tyr-Lys-Phe-Leu-
—/— T19 —/— T20 —/—

140 150 160
Ala-Gln-His-Ala-Leu-Arg-Trp-Asp-Asp-Asp-Tyr-Val-Pro-(His,Pro)-Val-Ile-Arg-Ile-Val-Glu-Pro-Ser-Trp-Val-
— T21 —/— T22 —/— T23 —

170 180
Gly-Ser-Asn-Asn-Glu-Tyr-Arg-Ile-Ser-Leu-Ala-Lys-Lys-Gly-Gly-Gly-Cys-Pro-Ile-Met-Asn-Leu-(Asx,Ser,Glx,
—/— T24 —/— T25 —/—

190 200 210
Tyr)-Thr-Asx-Ser-Phe-Glx-Glx-Phe-Ile-Asp-Arg-Val-Ile-Trp-Glu-Asn-Phe-Tyr-Lys-Pro-Ile-Val-Tyr-Ile-Gly-
— T32 —/— T27 —/—

220 230
Thr-Asp-Ser-Thr-Glu-Glu-Glu-Glu-Leu-Ile-Leu-Glu-Val-Ser-Leu-Val-Phe-Lys-Val-Lys-Glu-Phe-Ala-Pro-Asp-
— T28 —/— T29 —/— T30 —/—

240
Ala-Pro-Leu-Phe-Thr-Gly-Pro-Ala-Tyr
— T31 —/

Fig. 3. Reconstruction pattern of *G. mellonella* NPV polyhedrin polypeptide chain

Met-...

15 20

...-Thr-(Asp,Asn,Val,Tyr$_2$)-Lys-Phe-Tyr-Lys-(Ser,Val$_2$,Ile,Leu$_2$)-

T1 T2 T3

30 40 50

Lys-Asn-Ala-Lys-Lys-Lys-Glu-His-Leu-Ile-(Lys,His,Glu$_2$,Ile,Leu)-Lys-Glu-(Glx$_2$,Leu)-Lys-(Asp,Glu,Pro$_3$,

T4 T5 T6 T7

60 70 80

Gly$_2$,Ala,Val$_2$,Leu,Tyr,Phe$_2$)-Lys-Asn-Gln-Lys-Leu-Thr-Leu-Phe-Lys-Glu-Val-Arg-Asn-Val-Lys-Pro-Asp-Thr-

T8 T9 T10 T11 T12

90 100

Met-Lys-(His,Val$_2$,Asn,Trp)-Ser-Gly-Lys-Glu-Phe-Leu-Arg-Glu-Thr-Trp-Thr-Arg-Phe-Met-Glx-Asx-Ser-Phe-

T13 T14 T15

110 120 130

Pro-(Asx$_3$,Glx$_2$,Val$_3$,Ile$_2$)-Phe-(Lys,His,Arg$_2$,Asn,Thr,Gln,Pro$_2$,Ala$_2$,Leu)-Cys-Tyr-Arg-Phe-Leu-Ala-Gln-

T16 T17 T18 T19

140 150 160

His-Ala-Leu-Arg-Cys-(His,Asx$_2$,Glx,Pro$_2$,Val$_2$,Ile,Tyr)-Arg-Ile-Val-Glu-Pro-Asx-Tyr-Val-(Asx,Glx,Gly$_2$,

T20 T21 T22

170 180

Val,Tyr)-Arg-Ile-Ser-Leu-Ala-Lys-Arg-Gly-(Asn,Pro,Gly$_2$,Cys,Met,Ile,Leu)-Asx-Ser-Glx-Tyr-Asx-Asx-Ser-

T23 /T24/ T32 T25

190 200 210

(Ser,Glx$_2$,Ile,Phe$_2$)-Arg-Val-Ile-Trp-Glu-Asn-Phe-Tyr-Arg-(Asn$_2$,Val,Tyr)-Ile-Gly-Thr-Asx-(Ser,Glx$_5$,Ala,

T26 T27

220 230 240

Ile,Leu)-Leu-Leu-Ser-Leu-(Leu,Phe)-Lys-Val-Lys-Glu-Phe-Ala-Pro-Asx-Ile-Pro-(Asx,Ser,Glx,Gly,Val,Leu)-Tyr

T28

T29 T30 T31

Fig. 4. Reconstruction pattern of *A. segetum* NPV polyhedrin polypeptide chain

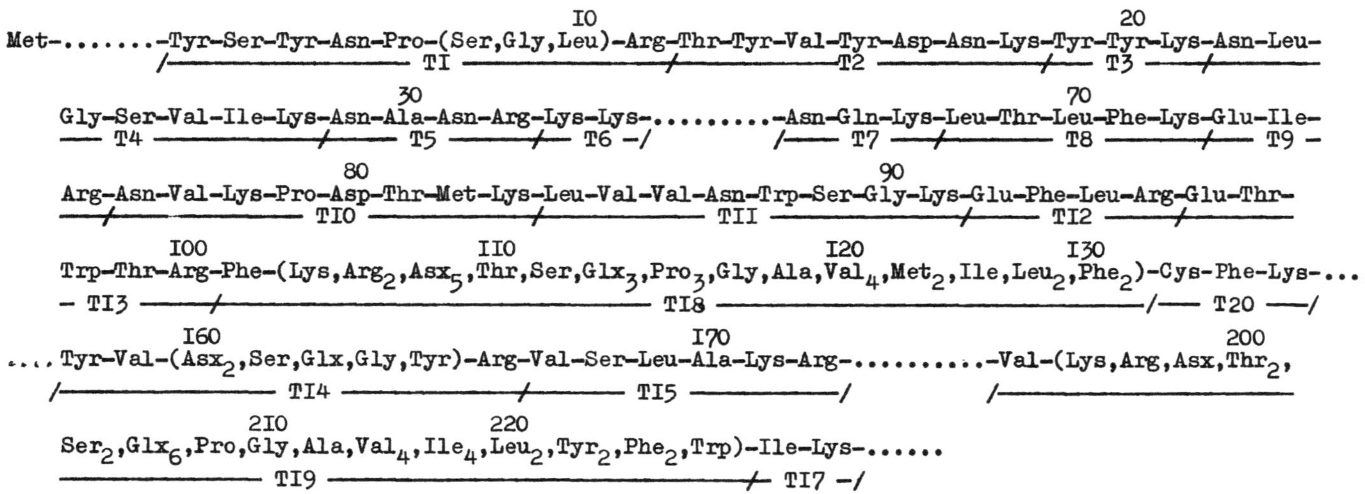

Fig. 5. Reconstruction pattern of the fragments of *M. brassicae* NPV polyhedrin polypeptide chain

cae (KOZLOV and SEREBRYANI 1985) NPV polyhedrins were fractionated and sequenced by the same methods as the standard protein. The reconstruction patterns of the polypeptide chain of the *P. dispar*, *G. mellonella*, and *A. segetum* NPV polyhedrins and polypeptide chain fragments of the *M. brassicae* NPV polyhedrin are shown in Figs. 2–5, respectively. The N-terminal methionine residue in *A. segetum* and *M. brassicae* NPV polyhedrins were determined starting with the intact proteins. The data in Fig. 2 show that the NPV polyhedrin sequence of *P. dispar* has a gap between position 194–199. No peptide could be located there as compared with *B. mori* NPV polyhedrin. One may suppose that the *P. dispar* polyhedrin gene could have a similar deletion. An alternative suggestion would be that an existing peptide had been lost during the isolation procedures. Not all the *A. segetum* and *M. brassicae* NPV polyhedrin peptides have been purified yet.

A comparison of peptide maps and amino acid compositions (KOZLOV et al. 1975b; 1985a; SUMMERS and SMITH 1975, 1976) shows that primary structures of polyhedrins and granulins differ significantly. Granulin was cleaved by trypsin and chymotrypsin to obtain overlapping peptides (KOZLOV et al. 1985b). The data in Table 1 give the composition and structure of the peptides obtained by us with reference to the related sequences in the *Trichoplusia ni* GV granulin according to AKIYOSHI et al. (1985).

It must be noted that granulin is poorly cleaved by trypsin and chymotrypsin. The majority of the protein remains uncleaved and maintains the form of large peptides. A minor part is digested to small peptides. Furthermore, trypsin reveals significant chymotryptic activity and chymotrypsin shows tryptic activity. As a result we obtained a few "bridge" peptides. The majority of the granulin peptides obtained differ from the corresponding polyhedrin peptides. Therefore, the complete granulin polypeptide chain cannot be reconstructed exclusively from tryptic and chymotryptic peptides. We reconstructed the sequences of some of the *A. segetum* GV granulin polypeptide chain fragments by comparing peptides listed in Table 1 with the sequence of several polyhedrins and *T. ni* GV granulin, which were derived from their DNA sequence (AKIYOSHI et al. 1985). Figure 6 shows the reconstruction pattern. Work on the complete amino acid sequence of *A. segetum* GV granulin is still in progress.

A comparison of polyhedrin and granulin amino acid sequences, studied in our laboratory, and of *T. ni* GV granulin (AKIYOSHI et al. 1985) with *Autographa californica* NPV polyhedrin sequences (HOOFT VAN IDDEKINGE et al. 1983) is shown in Fig. 7. The sequences were derived from the DNA sequences. The *A. segetum* GV granulin peptides listed in Table 1 have been aligned in a way that maximizes homology. We do not exclude the possibility of transposition of some *A. segetum* GV granulin peptides after the establishment of the complete amino acid sequence. Amino acids in brackets have been positioned preliminarily in order to maximize homology.

To determine the minimal number of genetic events that are required to generate a homologous set of the proteins shown in Fig. 7 we constructed a matrix showing amino acid differences (HOOD et al. 1975) (Table 2). For the calculation of the probable "degree of relatedness" of the proteins in question, we used only set A from the three amino acid homology sets proposed by

Table 1. Amino acid sequence and composition of some tryptic and chymotryptic peptides of *A. segetum* GV granulin and the possibly related *T. ni* GV granulin (AKIYOSHI et al. 1985)

Peptide	Sequence or composition	Related sequence
Ch44	Gly-Tyr	2–3
Ch22	Asn-Lys	4–5
T1	(Thr_2,Glx,Pro,Gly,Ala,Leu,Tyr_2)-Arg	6–15
Ch11	Ser-Leu-(Lys,Asx_2,Val,Ile)	16–22
Ch13	Lys-Gly-Tyr-(Lys,Asn,Ser,Gln,Gly,Ala,Ile)-Lys	22–32
T4	Asn-Val-Arg	33–35
Ch46	Lys,His,Arg_3,Glx,Pro,Ile,Leu_2	36–45
T18	Tyr-Ala-Pro-Leu-Lys	46–50
Ch47	Lys-Asx-Leu-Ala-Asx-Glx-Tyr	50–56
Ch31	Met-Val-(Asx,Thr,Glx,Pro)-Phe-Arg	57–64
Ch23	Gly-Pro-Gly-(Lys,Asn,Ser)-Lys	65–71
T6	Ile-Thr-Leu-Phe-Lys	72–76
T7	Glu-Ile-Arg	77–79
Ch12	Lys-Glu-Ile-Arg	76–79
T19	Arg	80
Ch20	Lys,Asx_2,Thr,Pro,Val,Met	81–87
Ch48	Lys-Leu-Ile-Xxx-Asx-Trp	87–92
Ch19	Ser-Gly-Lys-Glu-Phe	93–97
T8	Glu-Phe-Leu-Arg	96–99
Ch10	Leu-Arg-Glu-Thr-Trp	98–102
T9	Glu-Thr-Trp-Thr-Arg	100–104
Ch25	Phe-Ile-Glx-Asx-Glx-Phe-(Asx_2,Thr,Ser,Glx_2,Pro,Val,Met,Ile,Leu)-Tyr-Phe-Glu-Leu	105–125
Ch27	Asn,Leu	126–127
Ch9	Lys-Asp-Arg-(Lys,Arg,Cys,Ser,Glu,Ala,Ile,Tyr)	128–138
Ch18	Ser-Leu-(Arg,Tyr)	139–142
Ch5	Tyr-Ala-Leu-Gly-Ala-His-(His,Asx,Pro,Ala,Val,Tyr)-Asp-Val-Ile-Arg	142–157
T12	Ile-Thr-Glu-Pro-Ser-Phe-(Asx,Glx,Met,Leu)-(Asx,Glx,Pro)-Arg	158–171
T13	Ile-Asn-Leu-Ser-Lys	172–176
T15	Lys,His,Pro_2,Gly_3	177–183
Ch24	Thr-Cys-Leu-Gln	184–187
T17	Ser-Val-Tyr-Asx-(Asx,Glx)-Phe-(Asx,Phe_2)	188–197
Ch3	Phe-Asp-Glu-His-Leu-(Pro,Tyr,Trp)	198–205
Ch49	Phe-Phe	206–207
Ch21	Arg,Pro,Leu	208–210
Ch38	Val-Tyr	211–212
Ch1	(Asx,Gly,Met,Ile)-Ser-Ala-Glu-Ile-Glu-(Glu,Ile)-Met-(Glx,Ala,Val_3, Ile)-Phe	213–231
Ch14	Lys-Ile-Lys-Glu-Phe	232–236
Ch39	Ala-Pro-Asp-Val-Pro-Leu	237–242
Ch2	Phe-Thr-Gly-Pro-Ala-Tyr	243–248

HABER and KOSHLAND (1970). During the calculations we treated the gap in positions 198–203 of the *P. dispar* NPV polyhedrin as a deletion (KOZLOV et al. 1981a). The number of "correspondences" (identical and homologous amino acids) in the sequence comparison is adduced in Table 3. Obviously, the group of highly homologous proteins consists of polyhedrins, and we presume that

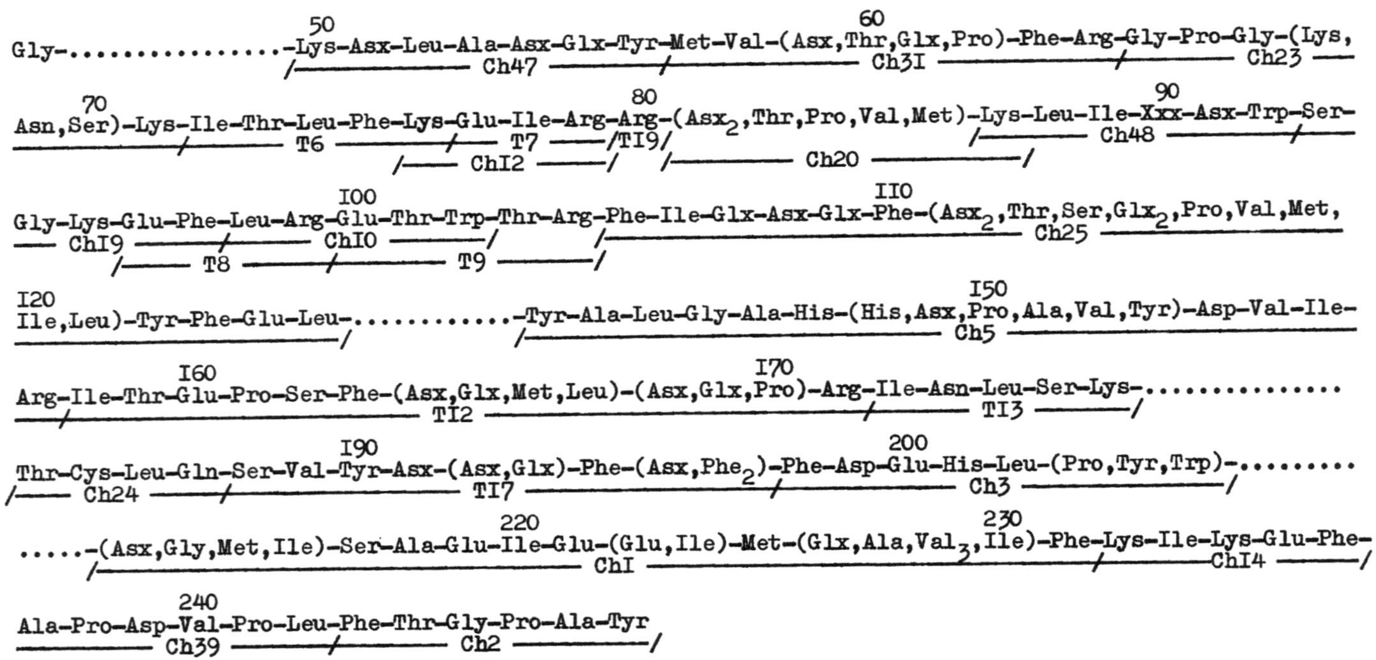

Fig. 6. Reconstruction pattern of the fragments of *A. segetum* GV granulin polypeptide chain. The amino acid sequence numeration is in agreement with *T. ni* GV granulin primary structure predicted by DNA sequencing (Akiyoshi et al. 1985). The unidentified amino acid residue is noted as Xxx in position 90

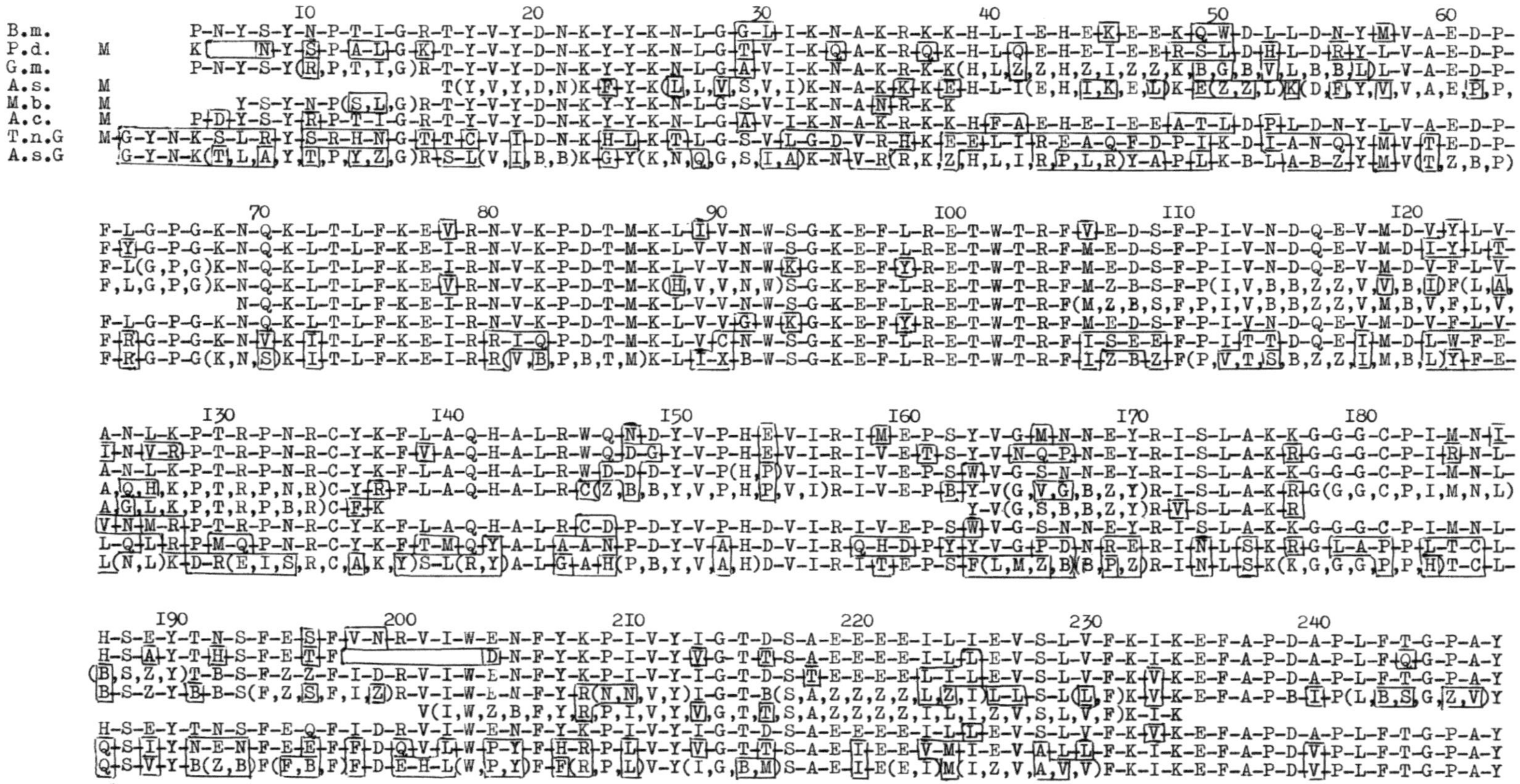

Fig. 7. Comparison of amino acid sequences of *B. mori* (*B.m.*), *P. dispar* (*P.d.*), *G. mellonella* (*G.m.*), *A. segetum* (*A.s.*), *M. brassicae* (*M.b.*), *A. californica* (*A.c.*) NPV polyhedrins, and *T. ni* (*T.n.G*) and *A. segetum* (*A.s.G*) granulins. *A. californica* NPV polyhedrin and *T. ni* GV granulin sequences are taken from Akiyoshi et al. (1985) and Hooft van Iddekinge et al. (1983) respectively. Framed *P. dispar* NPV sequence gaps reflect the supposed deletions in homologous genes. Framed amino acids reflect substitutions in homologous sequences. Gaps in *A. segetum* and *M. brassicae* NPV polyhedrin sequences mean that not all peptides have been isolated yet. The unidentified amino acid residue in *A. segetum* GV granulin position 90 is noted as X

Table 2. Amino acid difference matrix for polyhedrin and granulin sequences from Fig. 7

	A.c.	G.m.	B.m.	P.d.	A.s.	T.n.G
G.m.	19					
B.m.	35	32				
P.d.	53	54	54			
A.s.	58	52	52	71		
T.n.G	117	120	116	119	110	
A.s.G	116	116	108	119	118	74

Table 3. Numbers of "correspondences" (identical and homologous amino acid residues) between the polyhedrin and granulin sequences of Fig. 7

	A.c.	G.m.	B.m.	P.d.	A.s.	T.n.G	A.s.G
G.m.	229 (93%)	244					
B.m.	225 (92%)	229 (93%)	244				
P.d.	206 (84%)	206 (84%)	210 (85%)	244			
A.s.	193 (83%)	199 (85%)	202 (87%)	186 (80%)	233		
T.n.G	160 (65%)	156 (63%)	161 (65%)	155 (63%)	151 (64%)	247	
A.s.G	158 (64%)	157 (64%)	164 (66%)	158 (64%)	152 (64%)	205 (83%)	247

the granulins form the second group of highly homologous proteins. But granulins are less closely related to polyhedrins. The proteins within each group have a similar homology percentage, 80%–93%. The homology percentage of the *M. brassicae* NPV polyhedrin fragments to the same fragments of the other polyhedrins is 85%–93%. Homology percentage between the different groups of proteins is 63%–66%. The degree of relatedness of the *A. segetum* GV granulin to other inclusion body proteins may in fact turn out to be slightly lower for some of the peptide transpositions, once the complete amino acid sequence of the granulin has been established.

A striking peculiarity in the primary structure of the inclusion body protein is the presence of highly conserved and of variable amino acid sequences side by side. Sequence positions 65–122 (80%–100% homology) and 211–248 (82%–100% homology) are the highly conserved regions in the polypeptide chain. The functionally relevant residues such as Cys-135 and Cys-181/Cys-185 are strictly conserved. We consider the polyhedrin residue Cys-181 and the granulin residue Cys-185 functionally homologous. In all the examined proteins, map position 1–64 of the N-terminal amino acid sequence is the most variable region of the protein. The lowest "degree of relatedness" between granulins and polyhedrins is observed here, namely 39%–53%.

The stability in amino acid sequence among the six viral polyhedrins suggests that the stable regions on a polypeptide chain are the functionally relevant domains. It seems possible that in vivo functions of polyhedrins and granulins are different to a certain extent. It is interesting to note that GV have been found within Lepidoptera (Rohrmann et al. 1981).

3 Some Physicochemical Properties

We shall now shift to a brief consideration of some aspects which, in our view, are essential for a discussion of polyhedrin structure-function relationships.

3.1 Cleavage of Polyhedrin by Inclusion Body Protease

Baculovirus inclusion bodies contain a specific alkaline protease (IB protease) (EPPSTEIN and THOMA 1975; KOZLOV et al. 1975a; PAYNE and KOLMAKOFF 1978; TWEETEN et al. 1978; YAMAFUJI et al. 1960). There are a number of reports showing that the IB protease cleaves inclusion body protein during dissolution of the inclusion bodies under alkaline conditions (CIBULSKY et al. 1977; CRAWFORD and KALMAKOFF 1977; EPPSTEIN et al. 1975; HARRAP et al. 1977; KOZLOV et al. 1978; MCCARTHY and LIU 1976; SCHARNHORST and WEAVER 1980; SUMMERS and SMITH 1975; WOOD 1980). While studying the primary structure of *B. mori* NPV polyhedrin we attempted to determine the bonds which are cleaved by IB protease. It is shown by sodium dodecyl sulfate (SDS) polyacrylamid gel electrophoresis (PAGE) that *B. mori* NPV polyhedrin is cleaved by IB protease at pH 10.5 to a series of large fragments among which are major polypeptides with molecular weights of 22K, 19K, 10K, and 8K are (Fig. 8). The cleavage pattern is characteristic for each of the polyhedrins examined (KOZLOV et al. 1978). Only the quantitative correlations among the fragments vary. The *A. segetum* GV granulin obtained by dissolution of the granula is cleaved so extensively that SDS-PAGE reveals only one diffuse band in a molecular weight range from 8–10K (KOZLOV et al. 1985a). It is obvious that *A. segetum* GV granules contain a highly active protease or proteases.

In the course of studies on the N-terminal residues and tryptic peptides of the intact and of the IB protease-cleaved *B. mori* NPV polyhedrin, we determined which bonds were cleaved by IB protease. They are Trp-Ser (88–89), Tyr-Leu (118–119), Ala-Asn (121–122), Tyr-Thr (186–187), Phe-Val (193–194), Tyr-Lys (203–204), and Tyr-Ile (208–209). These are marked in Fig. 1 by arrows. From these data it appears that the IB protease has chymotrypsinlike activity. The same conclusion was reached earlier by EPPSTEIN and THOMA (1975), who studied the cleavage of synthetic substrates by IB protease.

We calculated the molecular weights of all possible fragments that could be derived from a polyhedrin polypeptide chain (28K) by IB protease cleavage. The 22K, 19K, 10K, and 8K major polypeptides (Fig. 8) correspond to polypeptide chain fragments 1–208, 89–244, 1–88, and 122–186. It is interesting to compare our data with those obtained by EPPSTEIN and THOMA (1977) and EPPSTEIN et al. (1975). These authors isolated the 9.4K N-terminal fragment of *T. ni* NPV polyhedrin which was degraded by IB protease. The molecular weight determinations for the 1–88, 10K N-terminal of the *B. mori* NPV polyhedrin are in agreement. It is surprising that the bonds cleaved by IB protease are conserved for all of the proteins in question. The only exception is the

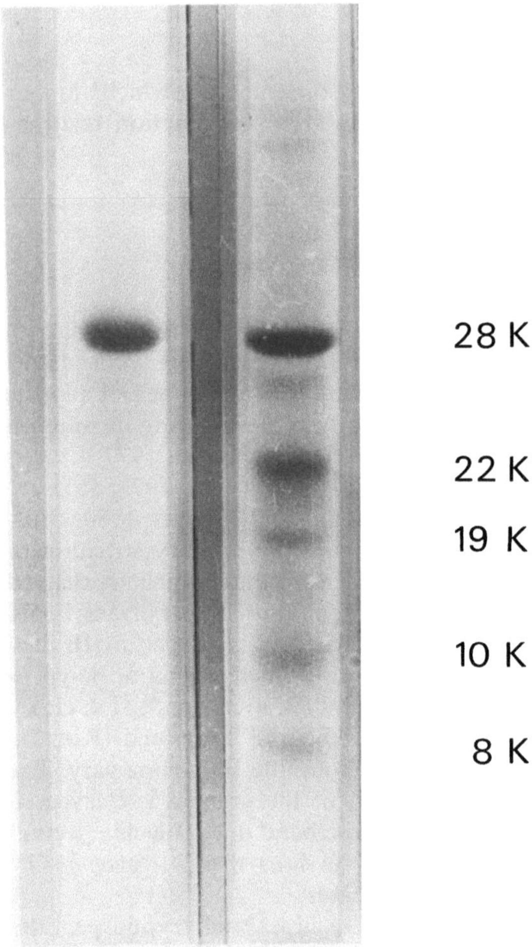

Fig. 8 a, b. Polyacrylamide gel electrophoresis patterns in tris-HCl buffer containing sodium dodecyl sulfate of the intact (**a**) and cleaved by IB protease (**b**) *B. mori* NPV polyhedrin

121–122 Ala-Asn bond. But this bond is not chymotrypsinspecific. Perhaps inclusion bodies contain proteases other than the chymotrypsinlike protease.

Although 40–50 polyhedrin bonds can be cleaved by chymotrypsin, only a limited number of these bonds is actually hydrolyzed by IB protease. *A. segetum* GV granule protease(s) is an exception (see above). This apparent discrepancy can be explained by conformational peculiarities and the quaternary structure of the protein under the conditions of cleavage at pH 10.5 (see also Sect. 3.2). We suggest that such specific cleavage of baculovirus polyhedrin by IB protease is not a random process but possesses a certain biological sense selected during the coevolution of baculoviruses and their hosts. Specific relationships during replication between proteases and virus proteins are a phenomenon observed in many viral families. Some aspects of this problem are elucidated in the review by HERSHKO and FRY (1975).

3.2 Association-Dissociation Processes and Secondary Structure

The extraordinary tendency to associate in solution is a characteristic feature of viral structural proteins. As a rule, the native subunit of a viral structural protein cannot be ascertained because the subunit may have arisen by dissolution of the protein under extreme conditions. This rule also holds for the baculovirus inclusion body proteins.

The inclusion body protein subunit is released only in concentrated acetic acid or in dilute alkaline solutions (KOZLOV et al. 1975a). We presume that the conformation of a polyhedrin molecule remains nearly native in solutions at pH values <10.5. This is confirmed by investigations on the secondary and quaternary structures of *B. mori* NPV polyhedrin (KOZLOV et al. 1981b). The least denaturing method, except for detergents, for polyhedrin preparation was described by BERGOLD (1947). Polyhedrin obtained by this method (pH 10.5) reveals by ultracentrifugation three oligomeric components, among which the major component 1 has a sedimentation coefficient of 13S (Fig. 9a, Table 4). As we have shown (KOZLOV et al. 1981b) by circular dichroism and optical rotation dispersion methods, such a polyhedrin contains 53% of the ordered conformations (17% α-helix and 36% β-structures). A reduction of pH from 10.5 to 7.5 leads to an increase in association: the new 34S component appears and the ratio among components 1–3 changes (Fig. 9b, Table 4). During this

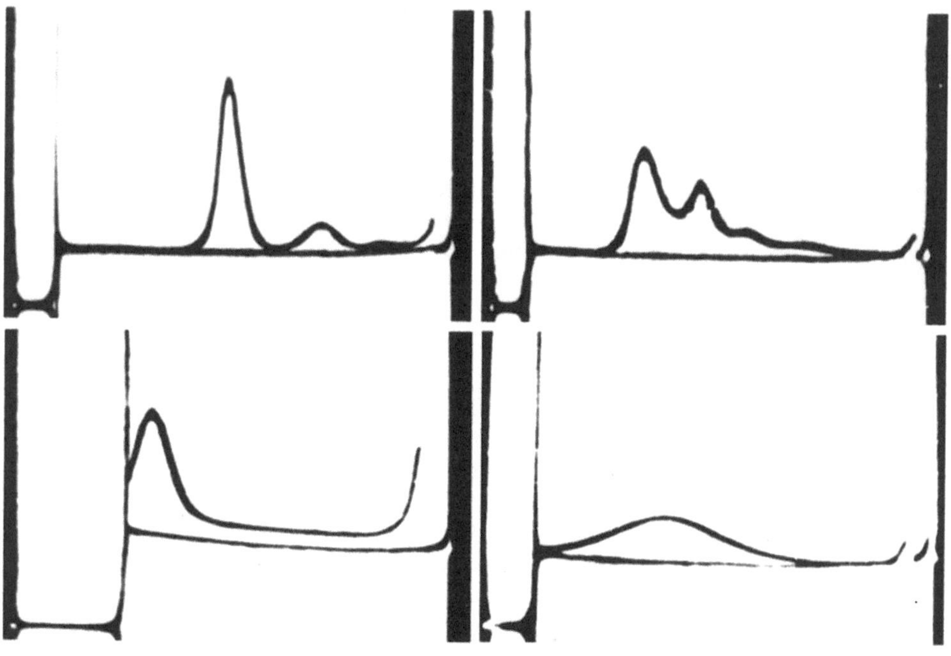

Fig. 9a–d. Sedimentation patterns of *B. mori* NPV polyhedrin at pH 10.5 (**a**), pH 7.5 (**b**), in 0.1 N NaOH (**c**), and after pH reduction from 12.9 to 7.5 (**d**). The pictures were taken at 50000 rpm after 45 min (**a**), 30 min (**b**), 65 min (**c**), and 15 min (**d**)

Table 4. Sedimentation coefficient (S) and component content (%) of *B. mori* NPV polyhedrin preparations of Fig. 9

Preparation characteristic	Conditions	Components 1 (S)	1 (%)	2 (S)	2 (%)	3 (S)	3 (%)	4 (S)	4 (%)	5 (S)	5 (%)
Intact protein of Fig. 8a	pH 10.5	13	70	18	20	23	10				
	pH 7.5	13	45	20	35	24	10	34	10		
	0.1 N NaOH									1.6	100
	67% CH_3COOH									1.5	100
Degradated protein of Fig. 8b	pH 10.5	13	70	19	20	23	10				
	pH 7.5	12	70	18	20	24	10				

Table 5. α-helix and β-structure content (%) of *B. mori* NPV polyhedrin preparation

Preparation characteristic	Conditions	α	β
Intact protein of Fig. 8a	pH 10.5	17	36
	pH 7.5	14	43
	0.1 N NaOH	11	18
	67% CH_3COOH	26	0
Degradated protein of Fig. 8b	pH 10.5	17	29
	pH 7.5	12	37

process, the total content in ordered conformations increases due to a reduction of the α-helical and an increase in the β-structure component (Table 5). Reduction of pH below 7.0 leads to precipitation of the protein. It seems that close to the isoelectric point the relative proportion of ordered structures continues to and reaches 60% (about 10% α-helix and 50% β-structures). Perhaps polyhedrin within native polyhedra contains the same quantity of ordered structure. We consider the 13S component to be the morphological unit of polyhedra. As was shown by SCHARNHORST and WEAVER (1980) by covalent cross linking, the 12S component in *Heliothis zea* NPV polyhedrin is dodecameric.

The process of association-dissociation (at pH values between 7.5 and 10.5) of the *B. mori* NPV polyhedrin is reversible but dissolution in 0.1 N NaOH or 67% acetic acid leads to irreversible changes. Under these conditions, polyhedrin reveals one component by ultracentrifugation (Fig. 9c), with a sedimentation coefficient of 1.6S in alkaline and of 1.5S in acid (Table 4). The contents of ordered conformations under these conditions are reduced to 29% and 26%, respectively (Table 5). The reduction of pH from 12.9 (0.1 N NaOH) to 7.5 by dialysis leads to the formation of a polydispersed system (Fig. 9d). However, even during this process, the tendency to an increase in ordered conformations is maintained. After the reduction of pH to 7.5, the polyhedrin contains 19%

of α-helical and 27% of β-structures. The association-dissociation process of polyhedrin cannot be studied under acid conditions because the polyhedrin precipitates during dilution of the acetic acid.

We have studied the influence of partial cleavage of *B. mori* NPV polyhedrin by IB protease on the association-dissociation process and on the establishment of secondary structures. The partially digested polyhedrin (Fig. 8b) reveals by ultracentrifugation at pH 10.5 the same sedimentation pattern as uncleaved protein (Fig. 9a; Table 4). Obviously, polyhedrin cleaved by IB protease is associated into particles at the same rate as uncleaved protein. But the contents of ordered conformations decreased, which accounts for a reduction in the content of β-structure (29% compared to 36% for uncleaved protein) (Table 5). It seems that the partial cleavage leads to some conformational changes which prevent polyhedrin association during reduction of pH from 10.5 to 7.6. This conclusion is confirmed by ultracentrifugation studies of the protein after reduction of pH from 10.5 to 7.5 by dialysis. The quantity of components in this protein and their ratio remain the same at pH 7.5 as at pH 10.5 (compare with data on the uncleaved protein, Table 4). Nevertheless, partially digested protein at reduced pH has the same tendency to a reduced α-helix content and to an increase in the β-structure content as uncleaved protein, although in general, the contents of ordered conformations are reduced to a certain extent (Table 5). It is obvious that cleavage by IB protease prevents polyhedrin from association into large particles. We have shown (KOZLOV and SEREBRYANI 1980a, b) that such a cleavage prevents polyhedrin association not only into large particles but also with virions. The virion release was considerably reduced when polyhedra were treated by protease inhibitors. This fact was noted by other authors also (TWEETEN et al. 1978; WOOD 1980). We suggest that the role of IB protease during infection is to make the virion release easier by converting the polyhedrin dissociation process from reversible to irreversible.

The secondary structure of *B. mori* NPV polyhedrin and its cyanogen bromide (BrCN) fragments in solutions containing 8 *M* urea and 1% SDS was investigated in our laboratory by KIBIREV et al. (1976). In 8 *M* urea solution polyhedrin is still associated but less than the "native" protein. A significant reduction of α-helical content was observed in 8 *M* urea but complete denaturation of the protein could not be detected. Authors assign polyhedrin to the group of proteins that acquire α-helical conformation under the influence of SDS, which accounts for destruction of β-structures. These authors suggest that the organization of the native globule and association of polyhedrin is reached mainly by hydrophobic interactions and by hydrogen bonds in a hydrophobic environment.

The secondary structure of some BrCN fragments was studied by KIBIREV et al. (1976). Most important appears to be the behavior of the C-terminal BrCN-V fragment in solution (see Fig. 1). This fragment contains 64 amino acid residues and has a molecular weight of 7.4K. It is the only one of the BrCN fragments that associates in solution, even in a solution of 50% formic acid (KIBIREV et al. 1976; RADAVSKI et al. 1972). The ability to associate is preserved in a solution of 1% SDS and is partly eliminated in a solution of 6 *M* urea (Fig. 10). As was shown by gel filtration, the BrCN-V fragment is

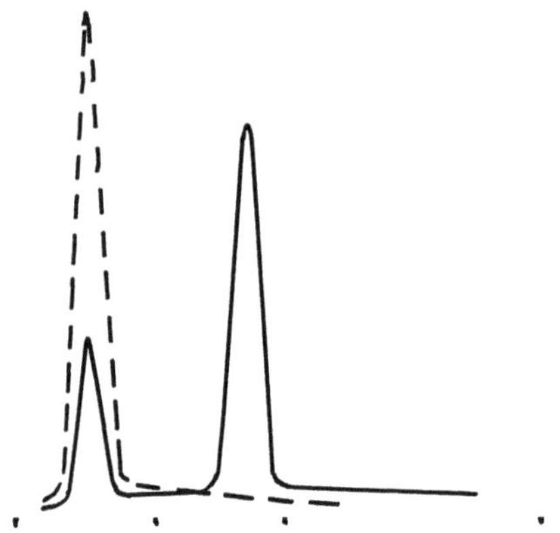

Fig. 10. Sephadex G-75 gel filtration of BrCN-V fragment (180–244) of *B. mori* NPV polyhedrin. Column was 1.8 × 90 cm, flow rate 30 ml/h, fraction volume 2 ml. Influents: ——, 6 *M* urea; -----, 1% sodium dodecyl sulfate

associated under these conditions into particles containing not less than 10 molecules. In solutions containig SDS and urea the behavior of the fragment is similar to that of the "native" protein. The content of α-helical structures increases and respectively the content of β-structures is reduced in a solution of 1% SDS. In urea solutions the decrease of association of the fragment proceeds parallel to a reduction of the content of α-helical structures. In a solution containing detergents, polyhedrin and its BrCN-V fragment have the same correlation between their association and the contents of ordered conformations. It may be supposed that the protein association of the BrCN-V fragment is maintained by hydrophobic interactions and hydrogen bonds in a hydrophobic environment.

On the base of these data it may be inferred that in solution the inclusion body proteins reveal strong protein–protein interactions leading to the formation of associates. There is a clearly expressed correlation between association and the secondary structure of the polyhedrin molecules. Protein–protein interactions are accompanied by the formation of β-structure regions and degradation of some α-helical regions.

3.3 Association with Polyhedral RNA

The presence of RNA in inclusion bodies was shown by many authors (Aizawa and Jida 1963; Cough 1971; Faulkner 1962; Himeno and Onodera 1969; Padhi and Chase 1976; Shapiro and Ignoffo 1971; Tarasevitch 1969) and by us (Kozlov et al. 1969, 1974). RNA containing fractions can be obtained after gel filtration of *B. mori* NPV polyhedrin solution through the Sephadex G-25 at pH 10.0 (Fig. 11). Pool 1 gives a weak RNA-specific test and its spectrum is proteinlike with a A_{260}/A_{280} ratio of 0.75. Pool 2 gives a positive test with *p*-bromophenylhydrazine and orcin, and its spectrum at pH 7.2 is nucleoproteinlike (Fig. 12) with a A_{260}/A_{280} ratio of 1.0. The presence of a polypeptide with a molecular weight of 28K in pool 2 was shown by SDS-PAGE analysis. After gel filtration at pH 11.0, all protein is detected in pool 1 (Fig. 13). This

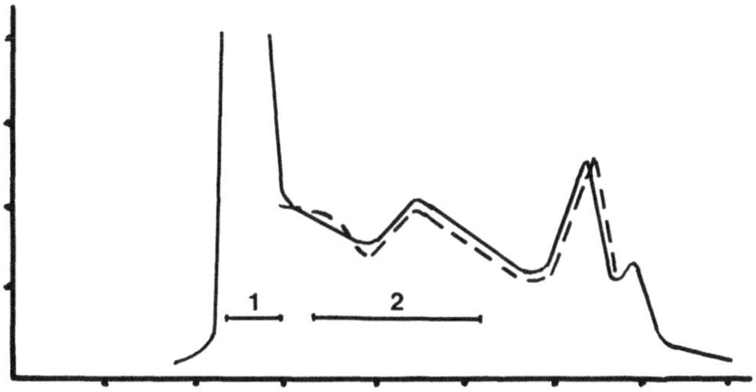

Fig. 11. Sephadex G-25 gel filtration in carbonate-hydrocarbonate buffer (pH 10.0) of *B. mori* NPV polyhedrin. Column was 3.5 × 40 cm, flow rate 30 ml/h, fraction volume 5 ml. —, absorbancy at 280 nm; -----, absorbancy at 260 nm

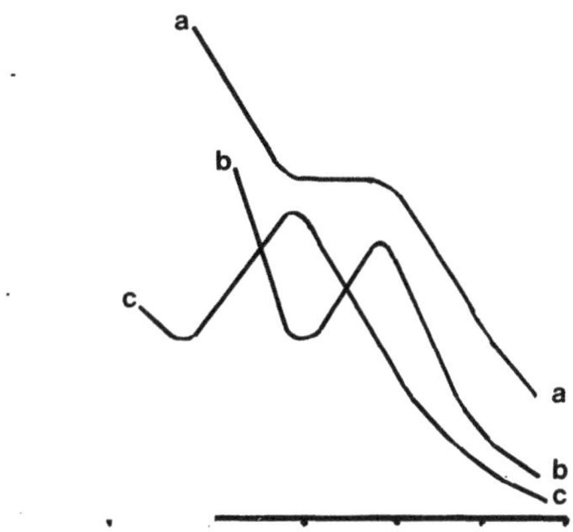

Fig. 12. Ultraviolet spectrum at pH 7.2 of pool 2 (Fig. 11) material (**a**), polyhedrin from pool 1 (Fig. 13) (**b**), and polyhedral RNA (**c**)

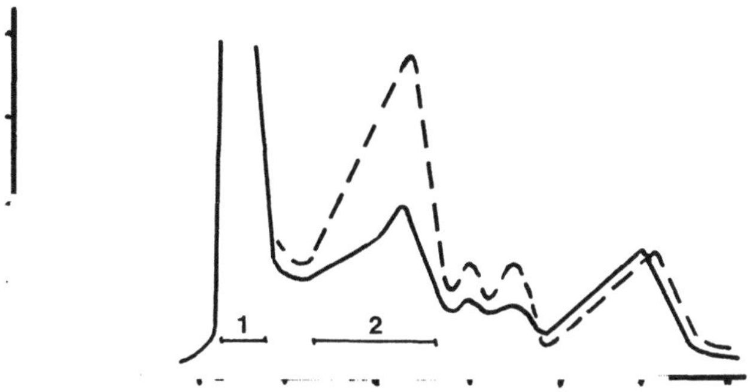

Fig. 13. Sephadex G-25 gel filtration in 0.01 M Na_2CO_3 and 0.05 M NaCl (pH 11.0) of *B. mori* NPV polyhedrin. For the other conditions, see Fig. 11

pool is devoid of RNA, and its spectrum is typical for protein with a A_{260}/A_{280} ratio of 0.60 (Fig. 12). Pool 2 (Fig. 13) contains only RNA or its fragments. Gel filtration of the polyhedrin solution at pH 7.2 after reversible reduction of the pH from 11.0 to 7.2 reveals the elution pattern presented in Fig. 11.

We have studied some physicochemical properties of the *B. mori* NPV RNA extracted from polyhedrin solution by the phenol method after dissolving the polyhedra at pH 10.5 and reducing the pH to 7.0 (KOZLOV et al. 1969). *B. mori* NPV RNA extracted by this method yields a positive test with *p*-bromo-

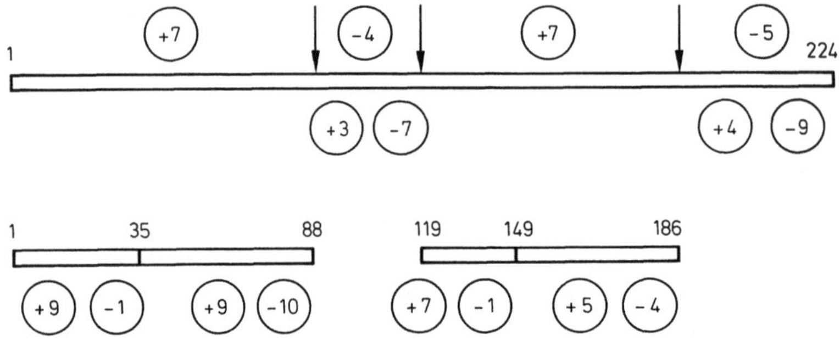

Fig. 14. The pattern of distribution of positively and negatively charged amino acid residues along the *B. mori* NPV polyhedrin polypeptide chain and its 1–88 and 119–186 fragments

phenylhydrazine and orcin, and its spectrum is characteristic for nucleic acid with a A_{260}/A_{280} ratio of 1.85 (Fig. 12). This RNA reacts with formaldehyde, leading to an increase in its A_{280} by 13% at 21° C and by 22% at 37° C.

Sephadex G-25 gel filtration at pH 10.0 of polyhedrin cleaved by IB protease (Fig. 8b) yields the same pattern as the uncleaved protein (Fig. 11). However, pool 2 reveals a diffuse band upon SDS-PAGE analysis with molecular weights ranging from 10–12K. It is conceivable that polyhedrin interacts by a specific fragment with polyhedra RNA. Gel filtration of *G. mellonella* NPV polyhedrin at pH 10.0 reveals the same pattern as *B. mori* NPV polyhedrin (Kozlov et al. 1974).

We explain the phenomenon described here for two different polyhedrins by the formation of a complex of polyhedrin with NPV RNA. It is suggested that lysine residues are essential for association with RNA because dissociation of the complex takes place at pH 11.0 (the lysine ε-NH_2-group dissociates at pH 10.5). It is thought that a RNA-polyhedrin complex is formed within the insect cells. It remains unknown whether RNA is included into polyhedra during their formation by haphazard inclusion, or whether this interaction is specific, and polyhedra RNA has a biological function.

4 Relations Between Structure and Function

In order to be able to evaluate the relationship between polyhedrin structure and function on the basis of our experimental data, we determined the distribution of amino acid residues along the polypeptide chain. For a theoretical analysis of the structure-function relationship we have applied the principle developed by Chippens et al. (1981). These authors, proceeding from the concept of biochemical universality, formulated a fundamental rule that aids finding structurally similar regions in peptide-protein ligands amongst those that differ in their origin and biological effect. These regions are the functionally relevant ones.

When they are in a 3-dimensional arrangement, they are responsible for the structure-function organization of a protein molecule.

The only known function of the inclusion body protein is the formation of the "supercapsid" around the virions which protects them from the environment while transferring infection from insect to insect. Typically the amino acid composition of viral capsid proteins corresponds to this function. This amino acid composition is characterized by the high content of hydrophobic amino acids, dicarbonic amino acids and their amides (50% and 23%, respectively). Thus, 73% of the amino acids in the inclusion body protein mainly take part in protein–protein interactions.

From the experimental data described in Sects. 3.2 and 3.3 we conclude that polyhedrin in solution associates with itself by protein–protein interactions, but also forms a complex with inclusion body RNA through protein–nucleic acid interactions. From published functions of polyhedrin on hemagglutinin activity (LITVINA 1975; REICHELDERFER 1974), ability to increase susceptibility of tissue cell culture to viral particles (VAUGHN and FAULKNER 1963), activation of the latent viruses (LONGWORTH and CUNNINGHAM 1968), antitumor potentiality (YAMAFUJI et al. 1971), and induction of nonspecific factor that inhibits macrophage migration (BENTON et al. 1973), it may be supposed that polyhedrin can also interact with the cell membrane. As chemically different amino acid residues are responsible for protein–protein, protein–nucleic and protein–membrane interactions, the polyhedrin polypeptide chain must possess some characteristic peculiarities in the distribution of functional amino acid residues. The role of hydrophobic interactions and hydrogen bonds in viral capsid stabilization was demonstrated (JACROT 1975; LAUFLER and STEVENS 1968). The basic (BRABURY et al. 1977; KAPER 1976; TSUGITA and HOSODA 1978) and aromatic (VEIKO et al. 1979) amino acid residues play an essential role in protein–nucleic acid interactions. The protein-peptide ligand interactions with membranes require specific active sites and structures often situated close to a cluster of basic amino acid residues (CHIPPENS et al. 1981).

We tried to perform a theoretical analysis of the distribution of the functionally important amino acid residues along the polypeptide chain. As polyhedrins are highly homologous proteins we analyzed only *B. mori* NPV polyhedrin. It seems possible that the distribution of acidic and basic amino acid residues may have characteristic features and correlates with the distribution of hydrophobic amino acid residues. For descriptive analyses, we divided the polypeptide chain into regions limited by some peptide bonds cleaved by IB protease (88–89, 118–119, and 186–187). Figure 14 shows the distribution of the acidic and basic residues along the whole polypeptide chain and along the isolated regions 1–88 and 119–186. It is apparent that the N-(1–88) and C-(119–244) terminal parts of the polypeptide chain are very similar in the distribution of charged residues. Within the terminal parts we can also distinguish C-terminal fragments (89–118 and 187–244) which are rich in acidic amino acid residues. These fragments are distinguished also by the high contents of hydrophobic residues (40%). In the positively charged regions (1–88 and 119–186), the distribution of basic amino acids is similar. The 1–36 and 119–149 N-terminal sequences have a high concentration of basic residues.

4.1 Region of Protein–Protein Interaction

With respect to protein–protein interactions, the 89–118 and 187–244 fragments are of great interest. From the data described in Sect. 2.2. it appears likely that the BrCN-V fragment (see Fig. 1) is responsible for the protein–protein interactions of the polypeptide chain. The BrCN-V fragment exceeds the 187–244 fragment in length only by six amino acid residues. Two tryptic core peptides, T27 and T16, which are insoluble under acidic conditions, are located here and in the 89–118 fragment, respectively. Their primary structures are very similar (Fig. 15). Both peptides contain clusters of hydrophobic amino acid residues which alternate with clusters of acidic amino acids and their amides. The physicochemical properties of the two peptides are also similar. The T27 peptide is insoluble at pH below 7.0, and T16 at pH below 5.0. In the tryptic peptide mixture, these peptides fail to migrate on paper electrophoresis at pH 6.5 (Fig. 15), though both of them contain clusters of acidic amino acid residues. But chymotryptic hydrolysis of the Leu-Val bond in the T27 peptide yields fragments that can be resolved by electrophoresis under the same conditions (Fig. 15). The T27 peptide itself remains at the start on paper electrophoresis at pH 6.5. It is obvious that the T27 peptide in solution forms an aggregate which dissociates after removal of the C-terminal sequence Val-Phe-Lys. We consider the T27 peptide to be the minimal region (residues 209–228) of the polypeptide chain that determines the protein–protein interactions of the BrCN-V fragment and hence of the 187–244 residue fragment of polyhedrin as well. Similarly, the T16 peptide (101–118) represents the analogous minimal region of the 89–118 residue fragment of polyhedrin.

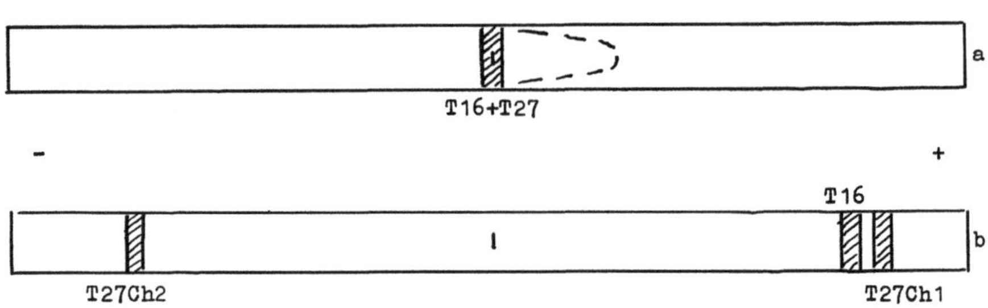

101 ... 118
Phe-Val-Glu-Asp-Ser-Phe-Pro-Ile-Val-Asn-Asp-Gln-Glu-Val-Met-Asp-Val-Tyr
/———— T16 ————/
Ch

209 ... 228
Ile-Gly-Thr-Asp-Ser-Ala-Glu-Glu-Glu-Glu-Ile-Leu-Ile-Glu-Val-Ser-Leu-Val-Phe-Lys
/———— T27 ————/
/———— T27Ch1 ————/— T27Ch2 —/

Fig. 15. Amino acid sequence of T27 and T16 *B. mori* NPV polyhedrin peptides and, at pH 6.5, paper electrophoresis of the peptide mixture (*a*) and their chymotryptic hydrolyzate (*b*)

Thus, we suppose that the 89–118 and 187–244 fragments of the polyhedrin polypeptide chain might determine the protein–protein interactions of the polyhedrin molecule. It seems possible that these fragments mainly contribute to the formation of polyhedra. Polyhedrin is known to associate firmly with the occluded virion membranes (KOZLOV and SEREBRYANI 1980a; WOOD 1980). We do not exclude the possibility of polyhedrin interaction with the virion membranes by the 89–118 or 187–244 residue fragments.

4.2 Region of Protein–Nucleic Acid Interaction

The data described in Sect. 3.3 show that the polyhedrin fragment presumably interacting with RNA has a molecular weight of not less than 10K. The N-terminal 1–88 residues possibly represent such a fragment. It is known that the majority of proteins binding to nucleic acids interact with their N-terminal sequences. To recognize the region responsible for protein–nucleic acid interactions, we compared the distribution of basic amino acid residues in the sequence of residues 1–37 in polyhedrin with that in the N-terminal part of other proteins which are known to bind to nucleic acids by their N-termini (Fig. 16). The characteristic feature of all the proteins examined is that basic amino acid residues are concentrated and especially clustered in the region in question. The possible role of lysine residues in polyhedrin interaction with polyhedral RNA was discussed above. The sequence of residues 3–20 of the polyhedrin is also marked by the presence of six tyrosine residues (40% of all tyrosine residues in the protein). The role of aromatic amino acid residues in protein–nucleic interactions is well known (VEIKO et al. (1979). Thus, we hypothesize that the sequence of residues 1–88 determines protein–nucleic acid interactions in the polyhedrin molecules.

4.3 Region of Protein–Membrane Interaction

To determine the region responsible for protein–membrane interactions, only a theoretical analysis was performed. In the 119–186 residue fragment we at-

```
             10        20          30
a   ..........R......K..K......K..KRKKH...
b   ..R.K..K...K...KRHRK..R.......K...RR..
c  KKR...H..R.R.H.R.....HKR.R..R.........
d   .......K..R..R...RK.KR..R
e   ........R.R.RRKK.K.R.....K...........
```

Fig. 16a–e. The distribution of basic amino acid residues in the N-terminal region of *B. mori* NPV polyhedrin (**a**), histone H4 (LISCHWE and SUNG 1977) (**b**), internal protein VII of adenovirus (LISCHWE and SUNG 1977) (**c**), bromovirus coat protein (TREMINE et al. 1977) (**d**), and C2 repressor of bacteriophage P22 (SAUER et al. 1981) (**e**). *K*, lysine; *R*, arginine; *H*, histidine. The cluster of basic amino acids is *underlined*

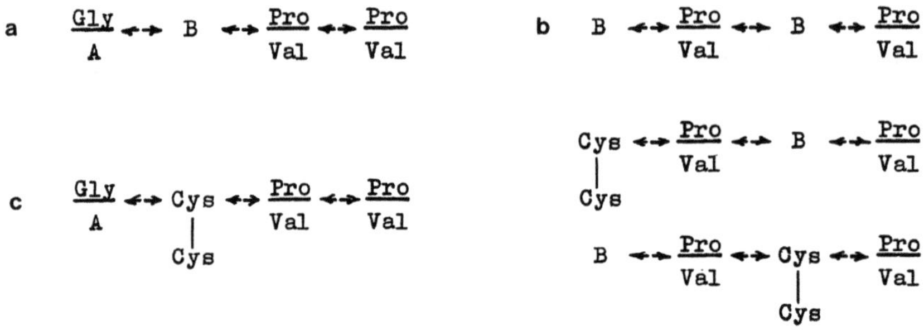

Fig. 17a–c. The primary structure of the type I (**a**) and type II (**b**) "common fragments" and cystine-containing analogues of "common fragments" (**c**) of peptide-protein ligands according to (CHIPPENS et al. 1981). *A*, amino acid with free carboxy group; *B*, basic amino acid or amide of dicarbonic amino acid; ↔, direction of the peptide chain acylation (direct or "retro")

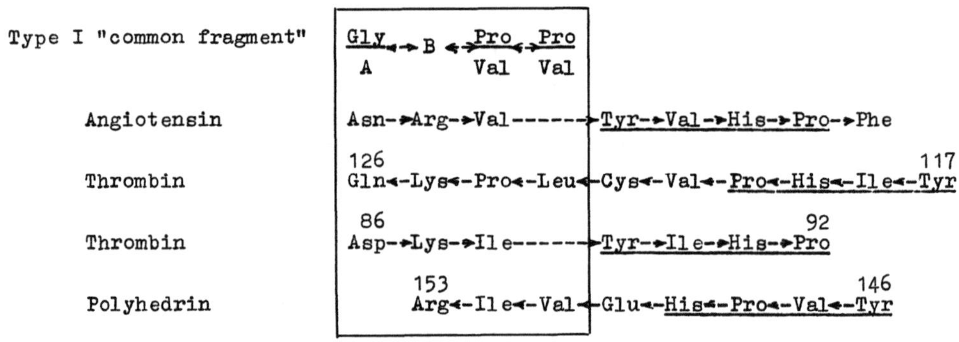

Fig. 18. Similarity between angiotensin sequences in *B. mori* NPV polyhedrin and thrombin polypeptide chains. The active site of angiotensin and homologous fragments of the proteins are *underlined*

tempted to find amino acid sequences similar to the "common fragments" and active sites of peptide–protein ligands, which are natural biological regulators. CHIPPENS et al. (1981) described two types of the "common fragments" in peptide-protein ligands. Their primary structure is shown in Fig. 17. The 119–186 residue region of polyhedrin contains a sequence that is strikingly similar to the primary structure of angiotensin (Fig. 18). This amino acid sequence was found by us in the thrombin polypeptide chain that is known to interact with receptors on the fibroblast surface (CUNNINGHAM et al. 1979) (Fig. 18). In the 119–186 residue region of polyhedrin, we have determined sequences that are very similar to the type II "common fragment" of peptide-protein ligands, to tuftsin, and to the cystine-containing type I "common fragment" analogue of peptide-protein ligands as well (Fig. 19). The conserved Cys-131 and Cys-177 residues in polyhedrin are situated in the 119–186 residue region. In this region IB protease cleaves nonspecific bonds: Ala-Asn (amino acids 121–122) in *B. mori* NPV polyhedrin and Gln-His (amino acids 137–138) in *G. mellonella* NPV polyhedrin (GUSAK et al. 1981; KOZLOV et al. 1981a). It is interesting to compare this finding with the fact that tuftsin and angiotensin

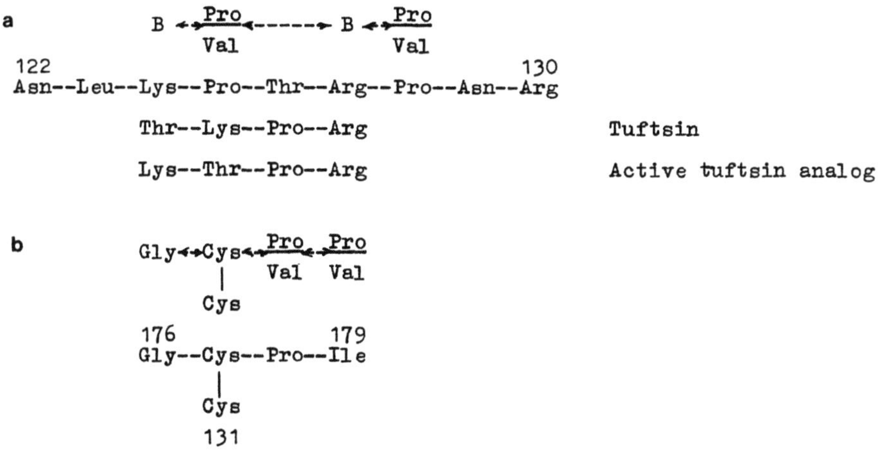

Fig. 19a, b. Amino acid sequences of *B. mori* NPV polyhedrin similar to the type II "common fragment" and tuftsin (**a**) and to the cystine-containing analogue of the type I "common fragment" (**b**)

are excised in vivo by sequential proteolysis from blood plasma globulins (CHIPPENS et al. 1981).

The similarity observed between some sequences in the 119–186 residue region of polyhedrin and the "common fragments" of peptide-protein ligands may be coincidental. However, we do not exclude the possibility that this region may possibly determine polyhedrin interactions with cell membranes.

4.4 Discussion

Taken together, the results of theoretical analyses of the primary structure and investigations of the physicochemical properties of the protein and its fragments suggest a hypothetical model of structure-function organization of the baculovirus polyhedrin polypeptide chain (Fig. 20). According to this model, four regions that determine protein–protein (residues 89–118 and 187–244), protein–nucleic acid (residues 1–88), and protein–membrane (residues 119–186) interactions can be depicted. On the one hand, the 180–244 residue C-terminal region, which is cleaved off by the IB protease, disturbs protein–protein interaction in order to enhance the liberation of virions. On the other hand, it can release fragments that are able to associate with RNAs or cellular membranes. It remains unknown whether such a cleavage occurs in insect cells. It is difficult to imagine that the hypothetical model described reflects the multifunctional properties of baculovirus polyhedrins. Such a suggestion is in contradiction with the data obtained from the exploration of genetically engineered deletion mutants (SMITH et al. 1983). The analysis of deletion mutants demonstrated that the polyhedrin gene was not essential for the production of infectious virus in tissue culture cells. But the infection process in insects is much more complicated than in cell culture. Little is known about the interaction between viral and cellular functions in

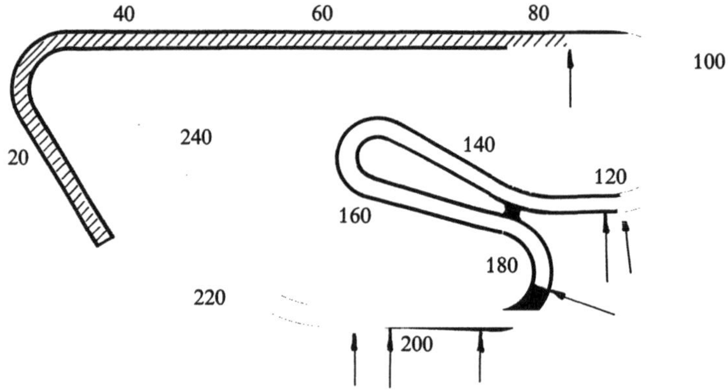

Fig. 20. The hypothetical model of the structure-function organization of the NPV polyhedrin polypeptide chain. The fragments responsible for protein–protein, protein–nucleic acid, and protein–membrane interactions are indicated as ■, ▨, and □, respectively. Arrows show the bonds that are cleaved by IB protease. It is supposed that IB protease cleaving the C-terminal fragment disturbs protein–protein interaction and enhances the liberation of virions. IB protease may also be capable of cutting out the functional fragments

different insect host organs. In addition, it is well known that silkworm and other larvae contain a latent virus. Insect polyhedrosis can be induced by various physical and chemical stress factors. Mirjuta et al. (1985) described the genome of *G. mellonella* NPV to be covalently bound to host cell DNA. It is conceivable that polyhedrin can be synthesized during infection by a polyhedrin-deletion mutant as a result of the induction of the polyhedrin function in the latent viral genome. For this reason we suggest that the role of polyhedrin during the interaction between virus and cell is not limited to the formation of inclusion bodies that protect virions from the environment and promote virus spread from insect to insect.

5 Conclusion

The determination of the primary structure of some baculovirus inclusion body proteins revealed clearly expressed peculiarities which were conserved during protein evolution. Our particular protein primary structure determination gave us the chance to study physicochemical properties not only of the intact protein but of its fragments as well. This allowed us to elucidate some aspects of the relationship between the primary structure and function, and to develop a hypothesis on the polypeptide chain structure-function organization of the baculovirus polyhedrin. The highly conserved primary structure (80%–90%) of polyhedrin suggests the possible importance of several of the polypeptide chain regions for protein functions. But it is difficult to apply this hypothesis to granulins. The granulins have the same high degree of homology to each other as the polyhedrins (83%), but they are homologous to polyhedrins to a lesser degree (63%–66%). It will be interesting to investigate whether the high degree of relatedness between granulins and polyhedrins will be maintained once the primary structures of additional granulins have been established.

The most interesting aspect in the evolution of inclusion body proteins of two baculovirus groups is the presence of prolonged regions of close homology (80%–100% of homology) and of variable regions in the polypeptide chain (minimum 39% of homology).

One of the surprising aspects of polyhedrin structure-function relationships is the interaction of inclusion body protein with IB protease. We assume that polyhedrin is the natural substrate of this protease. The high specificity of cleavage is one of the striking features in this relationship.

Further investigations on evolutional and structure-function peculiarities of inclusion body proteins will elucidate some aspects of the interactions between baculovirus, cell, and organism.

References

Aizawa K, Jida S (1963) Nucleic acids extracted from the virus polyhedra of the silkworm, *Bombyx mori* (Linnaeus). J Insect Pathol 5:334–348

Akiyoshi D, Chakerian R, Rohrmann GF, Nesson MH, Beaudreau GS (1985) Cloning and sequencing of the granulin gene from the *Trichoplusia ni* granulosis virus. Virology 141:328–332

Benton CV, Reichelderfer CF, Hetric FM (1973) Differentiation of *Trichoplusia ni* MEV and *Autographa californica* MEV by macrophage migration inhibition tests. J Invertebr Pathol 22:42–49

Bergold GH (1947) Die Isolierung des Polieder Virus und die Natur der Polieder. Z Naturforsch 2b:122–143

Brabury EM, Hjelm RP, Carpenter BC, Boldwin JB, Kneal CG, Hancock R (1977) Histones and chromatin structure. In: T'so POP (ed) The molecular biology of the mammalian genetic apparatus, vol 1. North-Holland, Amsterdam, pp 53–70

Chippens GI, Polevaja LK, Veretenikova NJ, Krikis AJ (1981) Structure and function of low-molecular peptides. Zinatne, Riga, p 328

Cibulsky RJ, Harper JD, Gudauskas RT (1977) Biochemical comparison of polyhedral protein from five nuclear polyhedrosis viruses infecting *Plusiine* larvae (Lepidoptera: Noctuidae). J Invertebr Pathol 29:182–191

Cough JA (1971) An enzootic nuclear polyhedrosis virus of pink shrimp: ultrastructure, prevalence, and enhancement. J Invertebr Pathol 24:311–331

Crawford AM, Kalmakoff J (1977) Effect of alkaline protease on the antigenic nature of *Wiseana* nuclear polyhedrosis virus polyhedrin protein. J Virol 24:412–415

Crook HE (1981) A comparison of the granulosis viruses from *Pieris brassicae* and *Pieris rapae*. Virology 115:173–181

Cunningham DD, Glenn KC, Backer JB, Simmer RL, Loww DA (1979) Mechanism of thrombin-stimulated cell division. J Supramol Struct 11:259–267

Eppstein DA, Thoma JA (1975) Alkaline protease associated with the matrix protein of a virus infecting the cabbage looper. Biochem Biophys Res Commun 62:478–484

Eppstein DA, Thoma JA (1977) Characterization and serology of the matrix protein from a nuclear polyhedrosis virus of *Trichoplusia ni* before and after degradation by an endogenous proteinase. Biochem J 167:321–332

Eppstein DA, Thoma JA, Scott HA, Young SY (1975) Degradation of matrix protein from a nuclear-polyhedrosis virus of *Trichoplusia ni* by an endogenous protease. Virology 67:591–594

Faulkner P (1962) Isolation and analysis of ribonucleic acid from inclusion bodies of the nuclear polyhedrosis of the silkworm. Virology 16:479–484

Gusak NM, Ovander MN, Drobot LB, Serebryani SB (1979) Determination of peptide structure by combination Dansyl-Edman method. In: Matsuka GC (ed) Methods of molecular biology. Naukova Dumka, Kiev, pp 142–154

Gusak NM, Kozlov EA, Ovander MN, Serebryani SB (1981) Tryptic peptides of inclusion body protein of nuclear polyhedrosis virus of *Galleria mellonella*. Bioorgan Chimija 7:996–1007

Gusak NM, Kozlov EA, Rodnin NV, Serebryani SB (1985) Structure of some tryptic peptides of *Agrotis segetum* nuclear polyhedrosis virus polyhedrin. Biopolimery Kletka 1:312–317

Haber JE, Koshland DE (1970) An evaluation of the relatedness of proteins based on comparison of amino acid sequences. J Mol Biol 50:617–639

Harrap KA, Payne CC, Robertson JS (1977) The properties of three baculoviruses from closely related hosts. Virology 79:14–31

Hershko A, Fry M (1975) Post-translational cleavage of polypeptide chains: role in assembly. Annu Rev Biochem 44:775–798

Himeno M, Onodera K (1969) RNA isolated from polyhedra of a nuclear polyhedrosis of the silkworm. J Invertebr Pathol 13:87–90

Hood LE, Wilson JN, Wood WB (1975) Molecular evolution. In: Molecular biology of eucaryotic cells, vol 1. Benjamin, London, pp 265–324

Hooft van Iddekinge BJL, Smith GE, Summers MD (1983) Nucleotide sequence of the polyhedrin gene of *Autographa californica* nuclear polyhedrosis virus. Virology 131:561–565

Jacrot B (1975) Studies on the assembly of a spherical plant virus. 2. The mechanism of protein aggregation and virus swelling. J Mol Biol 95:433–446

Kaper JM (1976) Molecular organization and stabilizing forces of simple RNA viruses. V. The role of lysyl residues in the stabilization of cucumber mosaic virus strain S. Virology 71:185–198

Kibirev VK, Radavski YL, Sucharenko NV, Serebryani SB (1976) Studies on the secondary structure of the polyhedral protein of the nuclear polyhedrosis virus of *Bombyx mori* and some of its fragments. Moleculjarn Biol 10:1272–1278

Kozlov EA, Serebryani SB (1980a) Virion polypeptides of the nuclear polyhedrosis virus of the silkworm *Bombyx mori*. Biochimija 45:130–136

Kozlov EA, Serebryani SB (1980b) Degradation of the silkworm polyhedrosis virus. Mikrobiol Zhurn 42:365–371

Kozlov EA, Serebryani SB (1985) Structure of some tryptic peptides of *Mamestra brassicae* nuclear polyhedrosis virus polyhedrin. Biopolimery Kletka 1:194–198

Kozlov EA, Soguljaeva VM, Levitina TL, Vereschak V, Serebryani SB (1969) Purification of polyhedral protein of nuclear polyhedrosis virus of silkworm and the processes of its association in solution. Biochimija 34:679–685

Kozlov EA, Levitina TL, Soguljaeva VM, Serebryani SB (1973) Separation and investigation of the amino acid composition of tryptic peptides from inclusion body protein of nuclear polyhedrosis virus of silkworm, *Bombyx mori*. Biochimija 38:1215–1220

Kozlov EA, Sidorova NM, Serebryani SB (1974) Obtaining and physico-chemical properties of polyhedral protein from nuclear polyhedrosis virus of *Galleria mellonella*. Biochimija 39:130–134

Kozlov EA, Sidorova NM, Serebryani SB (1975a) Proteolytic cleavage of polyhedral protein during dissolution of inclusion bodies of the nuclear polyhedrosis viruses of *Bombyx mori* and *Galleria mellonella* under alkaline conditions. J Invertebr Pathol 25:97–101

Kozlov EA, Levitina TL, Sidorova NM, Radavski YL, Serebryani SB (1975b) Comparative chemical studies of the polyhedral protein of the nuclear polyhedrosis viruses of *Bombyx mori* and *Galleria mellonella*. J Invertebr Pathol 25:103–107

Kozlov EA, Levitina TL, Gusak NM, Larionov GV, Veremejchenko SN, Serebryani SB (1978) Comparative biochemical studies of polyhedral proteins of nuclear polyhedrosis viruses. Biochimija 43:2189–2195

Kozlov EA, Levitina TL, Gusak NM, Ovander MN, Serebryani SB (1981a) Comparison of amino acid sequences of inclusion body proteins of nuclear polyhedrosis viruses of *Bombyx mori*, *Porthetria dispar*, and *Galleria mellonella*. Bioorgan Chimija 7:1008–1015

Kozlov EA, Lylo VV, Kibirev VK, Serebryani SB (1981b) Secondary structure and association-dissociation process in solution of *Bombyx mori* nuclear polyhedrosis virus inclusion body protein. In: Matsuka GC (ed) Moleculjarnaja biologija, vol 29. Naukova Dumka, Kiev, pp 42–48

Kozlov EA, Levitina TL, Gusak NM, Serebryani SB (1985a) Some physico-chemical properties of *Agrotis segetum* nuclear polyhedrosis virus inclusion body protein. Biopolimery Kletka 1:121–124

Kozlov EA, Levitina TL, Rodnin NM, Charchenko VM, Serebryani SB (1985b) Amino acid sequence of polypeptide chain fragments of *Agrotis segetum* granulosis virus granulin. Biopolimery Kletka 1:158–159

Laufler MA, Stevens CL (1968) Structure of the tobacco mosaic virus particles: polymerization of tobacco mosaic virus protein. Adv Virus Res 13:1–56
Levitina TL, Kozlov EA, Serebryani SB (1976) Tryptic peptides of polyhedral protein of nuclear polyhedrosis virus of *Bombyx mori*. Biochimija 41:228–236
Levitina TL, Kozlov EA, Ovander MN, Serebryani SB (1981) Tryptic peptides of inclusion body protein of nuclear polyhedrosis virus of *Porthetria dispar*. Bioorgan Chimija 7:985–995
Lischwe MA, Sung MT (1977) A histone-like protein from adenovirus chromatin. Nature 267:552–554
Litvina LA (1975) Haemagglutination test with insect viruses. Vopr Virusol 1:91–93
Longworth JF, Cunningham JC (1968) The activation of occult nuclear polyhedrosis viruses by foreign nuclear polyhedra. J Invertebr Pathol 10:361–367
McCarthy WJ, Liu SY (1976) Electrophoresis and serological characterization of *Porthetria dispar* polyhedron protein. J Invertebr Pathol 28:57–65
Mirjuta NYu, Strokovskaja LI, Kok IP, Skuratovskaja IN (1985) Finding of baculovirus genome covalent-bond with the cell DNA. Biopolimery Kletka 1:106–108
Padhi SB, Chase T (1976) Chemical characterization studies of *Porthetria dispar*. J Invertebr Pathol 28:137–142
Payne CC, Kolmakoff J (1978) Alkaline protease associated with virus particles of a nuclear polyhedrosis virus: assay, purification, and properties. J Virol 26:84–92
Radavski YuL, Kozlov EA, Serebryani SB (1972) Investigation of cyanogen bromide fragments of *Borrelinavirus bombycis* polyhedral protein. Chim Prirodn Soed 5:683–685
Radavski YuL, Kozlov EA, Serebryani SB (1973) Carboxypeptidase A action on *Borrelinavirus bombycis* polyhedral protein. Chim Prirodn Soed 4:569–570
Radavski YuL, Sucharenko NV, Kozlov EA, Serebryani SB (1976) Isolation and characterization of two cyanogen bromide fragments of polyhedral protein from nuclear polyhedrosis virus of *Bombyx mori*. Ukr Biochim Zh 48:706–711
Reichelderfer CF (1974) A haemagglutinating antigen from a nuclear polyhedrosis virus of *Spodoptera frugiperda*. J Invertebr Pathol 23:46–59
Rohrmann GF (1977) Characterization of N-polyhedrin of two baculovirus strains pathogenic for *Orgyia pseudotsugata*. Biochemistry 16:1631–1634
Rohrmann GF, Bailey TJ, Brimball B, Bocker RR, Beaudreau GS (1979) Tryptic peptide analysis and NH_2-terminal amino acid sequences of polyhedrins of two baculoviruses from *Orgyia pseudotsugata*. Proc Natl Acad Sci USA 76:4976–4980
Rohrmann GF, Pearson MN, Bailey TJ, Bocker RR, Beaudreau GS (1981) N-terminal polyhedrin sequences and occluded baculovirus evolution. J Mol Evol 17:329–333
Sauer RT, Pan J, Hopper P, Hehir K, Brown J, Poteete AR (1981) Primary structure of the phage P22 repressor and its gene C2. Biochemistry 20:3591–3598
Scharnhorst DW, Weaver RP (1980) A structural analysis of the matrix protein from the nuclear polyhedrosis virus of *Heliothis zea*. Virology 102:468–472
Shapiro M, Ignoffo CM (1971) Amino acid and nucleic acid analysis of inclusion bodies of the cotton boliworm, *Heliothis zea*, nucleopolyhedrosis virus. J Invertebr Pathol 18:154–155
Sidorova NM, Kozlov EA, Serebryani SB (1976) Fractionation and amino acid composition of acid-soluble trypsin peptides of inclusion body protein of moth (*Galleria mellonella*) nuclear polyhedron virus. Biochimija 41:781–786
Smith GE, Fraser MJ, Summers MD (1983) Molecular engineering of the *Autographa californica* nuclear polyhedrosis virus genome: deletion mutations within the polyhedrin gene. J Virol 46:584–593
Summers MD Smith GE (1975) *Trichoplusia ni* granulosis virus granulin: a phenol-soluble, phosphorylated protein. J Virol 16:1108–1116
Summers MD, Smith GE (1976) Comparative studies of baculovirus granulosis and polyhedrosis. Intervirology 6:168–180
Tarasevitch LM (1969) Nucleic acids of insect viruses and their hosts. Vopr Virusol 3:259–266
Tremine JH, Ronald WF, Agrawal HO (1977) Some tryptic peptides of bromovirus protein. Virology 83:404–412
Tsugita A, Hosoda J (1978) DNA binding site of the helix destabilizing protein gp32 from bacteriophage T4. J Mol Biol 122:255–258
Tweeten EA, Bulla LA Jr, Consigli RA (1978) Characterization of an alkaline protease associated with a granulosis virus of *Plodia interpunctella*. J Virol 26:702–711

Vaughn JL, Faulkner P (1963) Susceptibility of an insect tissue culture to infection by virus preparations of the nuclear polyhedrosis of the silkworm (*Bombyx mori* L.). Virology 20:484–489
Veiko NN, Gromova ES, Shabarova ZA (1979) The interaction of gene 5 protein of phage f1 with single- and double-stranded DNA and polynucleotides. Moleculjarn Biol 13:1136–1146
Wood HA (1980) Protease degradation of *Autographa californica* nuclear polyhedrosis virus proteins. Virology 103:392–399
Yamafuji K, Mukai J, Yoshihara F (1960) Deoxyribonuclease and protease in polyhedral viral particles. Enzymology 22:1–12
Yamafuji K, Ito M, Yoshihara F, Shinahara K (1971) Antitumour potentiality of polyhedral protein and its action on deoxyribonucleic acid. Z Krebsforsch 75:114–119

Subject Index

MIX
Papier aus verantwortungsvollen Quellen
Paper from responsible sources
FSC® C105338

If you have any concerns about our products,
you can contact us on
ProductSafety@springernature.com

In case Publisher is established outside the EU,
the EU authorized representative is:
Springer Nature Customer Service Center GmbH
Europaplatz 3, 69115 Heidelberg, Germany

Printed by Libri Plureos GmbH
in Hamburg, Germany